Abstract

This Recommendation specifies mechanisms for the generation of random bits using deterministic methods. The methods provided are based on either hash functions, block cipher algorithms or number theoretic problems.

KEY WORDS: deterministic random bit generator (DRBG); entropy; hash function; random number generator

Acknowledgements

The National Institute of Standards and Technology (NIST) gratefully acknowledges and appreciates contributions by Mike Boyle and Mary Baish from NSA for assistance in the development of this Recommendation. NIST also thanks the many contributions by the public and private sectors.

NIST Special Publication 800-90A

Recommendation for Random Number Generation Using Deterministic Random Bit Generators

Elaine Barker and John Kelsey

Computer Security Division
Information Technology Laboratory

COMPUTER SECURITY

January 2012

U.S. Department of Commerce
John Bryson, Secretary

National Institute of Standards and Technology
Patrick Gallagher, Director Under Secretary
of Commerce for Standards and Technology

NIST National Institute of Standards and Technology • Technology Administration • U.S. Department of Commerce

Table of Contents

Random Number Generation Using Deterministic Random Bit Generators

1 Authority

This publication has been developed by the National Institute of Standards and Technology (NIST) in furtherance of its statutory responsibilities under the Federal Information Security Management Act (FISMA) of 2002, Public Law 107-347.

NIST is responsible for developing standards and guidelines, including minimum requirements, for providing adequate information security for all agency operations and assets, but such standards and guidelines **shall not** apply to national security systems.

This recommendation has been prepared for use by Federal agencies. It may be used by nongovernmental organizations on a voluntary basis and is not subject to copyright. (Attribution would be appreciated by NIST.)

Nothing in this Recommendation should be taken to contradict standards and guidelines made mandatory and binding on federal agencies by the Secretary of Commerce under statutory authority. Nor should this Recommendation be interpreted as altering or superseding the existing authorities of the Secretary of Commerce, Director of the OMB, or any other federal official.

Conformance testing for implementations of this Recommendation will be conducted within the framework of the Cryptographic Module Validation Program (CMVP) and the Cryptographic Algorithm Validation Program(CAVP). The requirements of this Recommendation are indicated by the word "**shall**." Some of these requirements may be out-of-scope for CMVP or CAVP validation testing, and thus are the responsibility of entities using, implementing, installing or configuring applications that incorporate this Recommendation.

2 Introduction

This Recommendation specifies techniques for the generation of random bits that may then be used directly or converted to random numbers when random values are required by applications using cryptography.

There are two fundamentally different strategies for generating random bits. One strategy is to produce bits non-deterministically, where every bit of output is based on a physical process that is unpredictable; this class of random bit generators (RBGs) is commonly known as non-deterministic random bit generators (NRBGs)[1]. The other strategy is to compute bits deterministically using an algorithm; this class of RBGs is known as Deterministic Random Bit Generators (DRBGs)[2].

A DRBG is based on a DRBG mechanism as specified in this Recommendation and includes a source of entropy input. A DRBG mechanism uses an algorithm (i.e., a DRBG algorithm) that produces a sequence of bits from an initial value that is determined by a seed that is determined from the entropy input. Once the seed is provided and the initial value is determined, the DRBG is said to be instantiated and may be used to produce output. Because of the deterministic nature of the process, a DRBG is said to produce pseudorandom bits, rather than random bits. The seed used to instantiate the DRBG must contain sufficient entropy to provide an assurance of randomness. If the seed is kept secret, and the algorithm is well designed, the bits output by the DRBG will be unpredictable, up to the instantiated security strength of the DRBG.

The security provided by an RBG that uses a DRBG mechanism is a system implementation issue; both the DRBG mechanism and its source of entropy input must be considered when determining whether the RBG is appropriate for use by consuming applications.

---

[1] NRBGs have also been called True Random Number (or Bit) Generators or Hardware Random Number Generators.

[2] DRBGS have also been called Pseudorandom Bit Generators.

3 Scope

This Recommendation includes:

1. Requirements for the use of DRBG mechanisms,

2. Specifications for DRBG mechanisms that use hash functions, block ciphers and number theoretic problems,

3. Implementation issues, and

4. Assurance considerations.

This Recommendation specifies several diverse DRBG mechanisms, all of which provided acceptable security when this Recommendation was published. However, in the event that new attacks are found on a particular class of DRBG mechanisms, a diversity of **approved** mechanisms will allow a timely transition to a different class of DRBG mechanism.

Random number generation does not require interoperability between two entities, e.g., communicating entities may use different DRBG mechanisms without affecting their ability to communicate. Therefore, an entity may choose a single, appropriate DRBG mechanism for their consuming applications; see Annex G for a discussion of DRBG mechanism selection.

The precise structure, design and development of a random bit generator is outside the scope of this document.

NIST Special Publication (SP) 800-90B [SP 800-90B] provides guidance on designing and validating entropy sources. SP 800-90C [SP 800-90C] provides guidance on the construction of an RBG from a source of entropy input and an **approved** DRBG mechanism from this document (i.e., SP 800-90A).

4   Terms and Definitions

Algorithm	A clearly specified mathematical process for computation; a set of rules that, if followed, will give a prescribed result.
Approved	FIPS-**approved**, NIST-Recommended and/or validated by the Cryptographic Algorithm Validation Program (CAVP).
Approved entropy source	An entropy source that has been validated as conforming to SP 800-90B.
Backtracking Resistance	Backtracking resistance is provided relative to time T if there is assurance that an adversary who has knowledge of the internal state of the DRBG at some time subsequent to time T would be unable to distinguish between observations of ideal random bitstrings and (previously unseen) bitstrings that were output by the DRBG prior to time T. The complementary assurance is called Prediction Resistance.
Biased	A value that is chosen from a sample space is said to be biased if one value is more likely to be chosen than another value. Contrast with unbiased.
Bitstring	A bitstring is an ordered sequence of 0's and 1's. The leftmost bit is the most significant bit of the string and is the newest bit generated. The rightmost bit is the least significant bit of the string.
Bitwise Exclusive-Or	An operation on two bitstrings of equal length that combines corresponding bits of each bitstring using an exclusive-or operation.
Block Cipher	A symmetric key cryptographic algorithm that transforms a block of information at a time using a cryptographic key. For a block cipher algorithm, the length of the input block is the same as the length of the output block.
Consuming Application	The application (including middleware) that uses random numbers or bits obtained from an **approved** random bit generator.
Cryptographic Key (Key)	A parameter that determines the operation of a cryptographic function such as: 1. The transformation from plaintext to ciphertext and vice versa, 2. The generation of keying material,

	3. A digital signature computation or verification.
Deterministic Algorithm	An algorithm that, given the same inputs, always produces the same outputs.
Deterministic Random Bit Generator (DRBG)	An RBG that includes a DRBG mechanism and (at least initially) has access to a source of entropy input. The DRBG produces a sequence of bits from a secret initial value called a seed, along with other possible inputs. A DRBG is often called a Pseudorandom Number (or Bit) Generator.
DRBG Mechanism	The portion of an RBG that includes the functions necessary to instantiate and uninstantiate the RBG, generate pseudorandom bits, (optionally) reseed the RBG and test the health of the the DRBG mechanism.
DRBG Mechanism Boundary	A conceptual boundary that is used to explain the operations of a DRBG mechanism and its interaction with and relation to other processes.
Entropy	A measure of the disorder, randomness or variability in a closed system. Min-entropy is the measure used in this Recommendation.
Entropy Input	An input bitstring that provides an assessed minimum amount of unpredictability for a DRBG mechanism. (See min-entropy.)
Entropy Source	A process or mechanism that produces unpredictable digital data. See SP 800-90B. Contrast with the Source of Entropy Input.
Equivalent Process	Two processes are equivalent if, when the same values are input to each process, the same output is produced.
Exclusive-or	A mathematical operation; the symbol $\oplus$, defined as: $0 \oplus 0 = 0$ $0 \oplus 1 = 1$ $1 \oplus 0 = 1$ $1 \oplus 1 = 0$. Equivalent to binary addition without carry.
Fresh Entropy	A bitstring output from a source of entropy input for which there is a negligible probability that it has been previously output by the source and a negligible probability that the bitstring has been previously used by the DRBG.

Full Entropy	For the purposes of this Recommendation, an n-bit string is said to have full entropy if that bitstring is estimated to contain at least $(1-\varepsilon)n$ bits of entropy, where $0 \leq \varepsilon \leq 2^{-64}$. A source of full-entropy bitstrings serves as a practical approximation to a source of ideal random bitstrings of the same length (see ideal random sequence).
Hash Function	A (mathematical) function that maps values from a large (possibly very large) domain into a smaller range. The function satisfies the following properties: 1. (One-way) It is computationally infeasible to find any input that maps to any pre-specified output; 2. (Collision free) It is computationally infeasible to find any two distinct inputs that map to the same output.
Health Testing	Testing within an implementation immediately prior to or during normal operation to determine that the implementation continues to perform as implemented and as validated
Ideal Random Bitstring	See Ideal Random Sequence.
Ideal Random Sequence	Each bit of an ideal random sequence is unpredictable and unbiased, with a value that is independent of the values of the other bits in the sequence. Prior to the observation of the sequence, the value of each bit is equally likely to be 0 or 1, and, the probability that a particular bit will have a particular value is unaffected by knowledge of the values of any or all of the other bits. An ideal random sequence of n bits contains n bits of entropy.
Implementation	An implementation of an RBG is a cryptographic device or portion of a cryptographic device that is the physical embodiment of the RBG design, for example, some code running on a computing platform.
Implementation Testing for Validation	Testing by an independent and accredited party to ensure that an implementation of this Recommendation conforms to the specifications of this Recommendation.
Instantiation of an RBG	An instantiation of an RBG is a specific, logically independent, initialized RBG. One instantiation is distinguished from another by a "handle" (e.g., an identifying number).
Internal State	The collection of stored information about a DRBG instantiation. This can include both secret and non-secret

	information.
Key	See Cryptographic Key.
Min-entropy	The *min-entropy* (in bits) of a random variable X is the largest value m having the property that each observation of X provides at least m bits of information (i.e., the min-entropy of X is the greatest lower bound for the information content of potential observations of X). The min-entropy of a random variable is a lower bound on its entropy. The precise formulation for min-entropy is $-(\log_2 \max p_i)$ for a discrete distribution having probabilities $p_1,\dots, p_n$. Min-entropy is often used as a worst-case measure of the unpredictability of a random variable. Also see SP 800-90B.
Non-Deterministic Random Bit Generator (Non-deterministic RBG) (NRBG)	An RBG that (when working properly) produces outputs that have full entropy. Contrast with a DRBG. Other names for non-deterministic RBGs are True Random Number (or Bit) Generators and, simply, Random Number (or Bit) Generators.
Nonce	A time-varying value that has at most a negligible chance of repeating, e.g., a random value that is generated anew for each use, a timestamp, a sequence number, or some combination of these.
Personalization String	An optional string of bits that is combined with a secret entropy input and (possibly) a nonce to produce a seed.
Prediction Resistance	Prediction resistance is provided relative to time T if there is assurance that an adversary who has knowledge of the internal state of the DRBG at some time prior to T would be unable to distinguish between observations of ideal random bitstrings and bitstrings output by the DRBG at or subsequent to time T. The complementary assurance is called Backtracking Resistance.
Pseudorandom	A process (or data produced by a process) is said to be pseudorandom when the outcome is deterministic, yet also effectively random, as long as the internal action of the process is hidden from observation. For cryptographic purposes, "effectively" means "within the limits of the intended cryptographic strength."
Pseudorandom Number Generator	See Deterministic Random Bit Generator.
Public Key	In an asymmetric (public) key cryptosystem, that key of an

	entity's key pair that is publicly known.
Public Key Pair	In an asymmetric (public) key cryptosystem, the public key and associated private key.
Random Number	For the purposes of this Recommendation, a value in a set that has an equal probability of being selected from the total population of possibilities and, hence, is unpredictable. A random number is an instance of an unbiased random variable, that is, the output produced by a uniformly distributed random process.
Random Bit Generator (RBG)	A device or algorithm that outputs a sequence of binary bits that appears to be statistically independent and unbiased. An RBG is either a DRBG or an NRBG.
Reseed	To acquire additional bits that will affect the internal state of the DRBG mechanism.
Secure Channel	A path for transferring data between two entities or components that ensures confidentiality, integrity and replay protection, as well as mutual authentication between the entities or components. The secure channel may be provided using cryptographic, physical or procedural methods, or a combination thereof.
Security Strength	A number associated with the amount of work (that is, the number of operations of some sort) that is required to break a cryptographic algorithm or system in some way. In this Recommendation, the security strength is specified in bits and is a specific value from the set {112, 128, 192, 256}. If the security strength associated with an algorithm or system is S bits, then it is expected that (roughly) 2^S basic operations are required to break it.
Seed	Noun : A string of bits that is used as input to a DRBG mechanism. The seed will determine a portion of the internal state of the DRBG, and its entropy must be sufficient to support the security strength of the DRBG. Verb : To acquire bits with sufficient entropy for the desired security strength. These bits will be used as input to a DRBG mechanism to determine a portion of the initial internal state. Also see reseed.
Seedlife	The length of the seed period.
Seed Period	The period of time between instantiating or reseeding a DRBG

	with one seed and reseeding that DRBG with another seed.
Sequence	An ordered set of quantities.
Shall	Used to indicate a requirement of this Recommendation.
Should	Used to indicate a highly desirable feature for a DRBG mechanism that is not necessarily required by this Recommendation.
Source of Entropy Input (SEI)	A component of a DRBG that outputs bitstrings that can be used as entropy input by a DRBG mechanism. An SEI may be an **approved** entropy source, an **approved** RBG employing an **approved** entropy source to obtain entropy input for its DRBG mechanism, or a nested chain of **approved** RBGs whose initial member employs an **approved** entropy source to obtain entropy input for its DRBG mechanism.
String	See Bitstring.
Unbiased	A value that is chosen from a sample space is said to be unbiased if all potential values have the same probability of being chosen. Contrast with biased.
Unpredictable	In the context of random bit generation, an output bit is unpredictable if an adversary has only a negligible advantage (that is, essentially not much better than chance) in predicting it correctly.
Working State	A subset of the internal state that is used by a DRBG mechanism to produce pseudorandom bits at a given point in time. The working state (and thus, the internal state) is updated to the next state prior to producing another string of pseudorandom bits.

5 Symbols and Abbreviated Terms

The following abbreviations are used in this Recommendation:

Abbreviation	Meaning
AES	Advanced Encryption Standard, as specified in [FIPS197].
DRBG	Deterministic Random Bit Generator.
ECDLP	Elliptic Curve Discrete Logarithm Problem.
FIPS	Federal Information Processing Standard.
HMAC	Keyed-Hash Message Authentication Code, as specified in [FIP198].
NIST	National Institute of Standards and Technology
NRBG	Non-deterministic Random Bit Generator.
RBG	Random Bit Generator.
SP	NIST Special Publication
TDEA	Triple Data Encryption Algorithm, as specified in [SP800-67].

The following symbols are used in this Recommendation:

Symbol	Meaning
$+$	Addition
$\lceil X \rceil$	Ceiling: the smallest integer $r \geq X$. For example, $\lceil 5 \rceil = 5$, and $\lceil 5.5 \rceil = 6$.
$\lfloor X \rfloor$	Floor: The largest integer less than or equal to X. For example, $\lfloor 5 \rfloor = 5$, and $\lfloor 5.3 \rfloor = 5$.
$X \oplus Y$	Bitwise exclusive-or (also bitwise addition modulo 2) of two bitstrings X and Y of the same length.
$X \parallel Y$	Concatenation of two strings X and Y. X and Y are either both bitstrings, or both byte strings.
gcd (x, y)	The greatest common divisor of the integers x and y.
len (a)	The length in bits of string a.
x **mod** n	The unique remainder r (where $0 \leq r \leq n-1$) when integer x is divided by n. For example, 23 mod 7 = 2.
(switch symbol)	Used in a figure to illustrate a "switch" between sources of input.
$\{a_1, ...a_i\}$	The internal state of the DRBG at a point in time. The types and number of the a_i depends on the specific DRBG mechanism.

Symbol	Meaning
0x*ab*	Hexadecimal notation that is used to define a byte (i.e., 8 bits) of information, where *a* and *b* each specify 4 bits of information and have values from the range {0, 1, 2,...F}. For example, 0xc6 is used to represent 11000110, where c is 1100, and 6 is 0110.
0^x	A string of *x* zero bits.

6 Document Organization

This Recommendation is organized as follows:

— Section 7 provides a functional model for a DRBG that uses a DRBG mechanism and discusses the major components of the DRBG mechanim.

— Section 8 provides concepts and general requirements for the implementation and use of a DRBG mechanism.

— Section 9 specifies the functions of a DRBG mechanism that are introduced in Section 8. These functions use the DRBG algorithms specified in Section 10.

— Section 10 specifies **approved** DRBG algorithms. Algorithms have been specified that are based on the hash functions specified in [FIPS 180], block cipher algorithms specified in [FIPS197] and [SP 800-67] (AES and TDEA, respectively), and a number theoretic problem that is expressed in elliptic curve technology.

— Section 11 addresses assurance issues for DRBG mechanisms, including documentation requirements, implementation validation and health testing,

This Recommendation also includes the following appendices:

— Appendix A specifies additional information that is specific to the DRBG mechanism based on elliptic curves.

— Appendix B provides conversion routines.

— Appendix C discusses security considerations when extracting bits using the DRBG mechanism based on elliptic curves.

— Appendix D provides example pseudocode for each DRBG mechanism. Examples of the values computed for the DRBGs using each **approved** cryptographic algorithm and key size are available at http://csrc.nist.gov/groups/ST/toolkit/examples.html under the entries for SP 800-90A.

— Appendix E provides a discussion on DRBG mechanism selection.

— Appendix F provides a list of **shall** statements that are contained in this document that are not validatable using NIST's validation programs. Rather, these requirements are the responsibility of entities using, implementing, installing or configuring applications or protocols that incorporate this Recommendation.

— Appendix G provides references.

— Appendix H provides a list of modifications to SP 800-90A since it was first published.

7   Functional Model of a DRBG

Figure 1 provides a functional model of a DRBG (i.e., one type of RBG). A DRBG uses a DRBG mechanism and a source of entropy input, and may, depending on the implementation of the DRBG mechanism, include a nonce source. The components of this model are discussed in the following subsections.

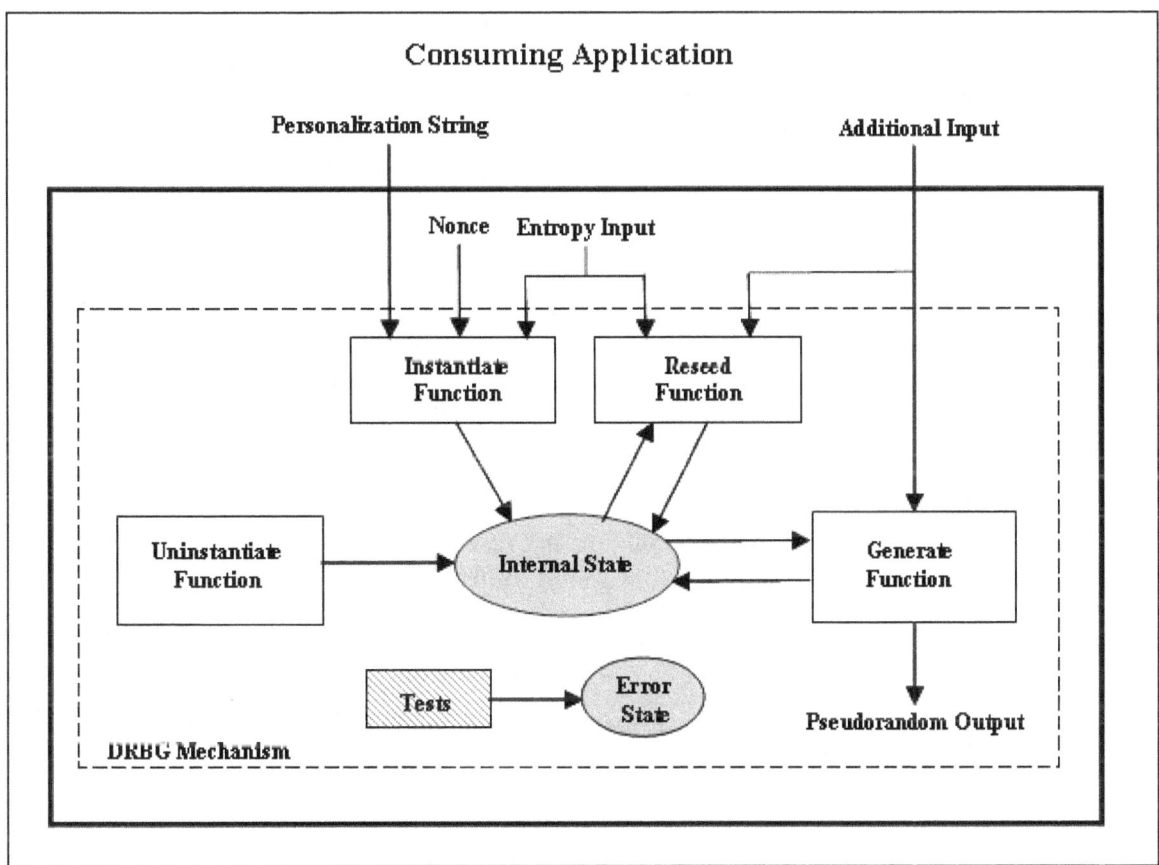

Figure 1: DRBG Functional Model

7.1   Entropy Input

The entropy input is provided to a DRBG mechanism for the seed (see Section 8.6). The entropy input and the seed **shall** be kept secret. The secrecy of this information provides the basis for the security of the DRBG. At a minimum, the entropy input **shall** provide the amount of entropy requested by the DRBG mechanism. Appropriate sources for the entropy input are discussed in Section 8.6.5.

Ideally, the entropy input will have full entropy; however, the DRBG mechanisms have been specified to allow for some bias in the entropy input by allowing the length of the entropy input to be longer than the required amount of entropy (expressed in bits). The entropy input can be defined to be a variable length (within limits), as well as fixed length.

In all cases, the DRBG mechanism expects that when entropy input is requested, the returned bitstring will contain at least the requested amount of entropy. Additional entropy beyond the amount requested is not required, but is desirable.

7.2   Other Inputs

Other information may be obtained by a DRBG mechanism as input. This information may or may not be required to be kept secret by a consuming application; however, the security of the DRBG itself does not rely on the secrecy of this information. The information **should** be checked for validity when possible; for example, if time is used as an input, the format and reasonableness of the time could be checked.

During DRBG instantiation, a nonce may be required, and if used, it is combined with the entropy input to create the initial DRBG seed. The nonce and its use are discussed in Sections 8.6.1 and 8.6.7.

This Recommendation strongly advises the insertion of a personalization string during DRBG instantiation; when used, the personalization string is combined with the entropy input bits and possibly a nonce to create the initial DRBG seed. The personalization string **shall** be unique for all instantiations of the same DRBG mechanism type (e.g., all instantiations of **HMAC_DRBG**). See Section 8.7.1 for additional discussion on personalization strings.

Additional input may also be provided during reseeding and when pseudorandom bits are requested. See Section 8.7.2 for a discussion of this input.

7.3   The Internal State

The internal state is the memory of the DRBG and consists of all of the parameters, variables and other stored values that the DRBG mechanism uses or acts upon. The internal state contains both administrative data (e.g., the security strength) and data that is acted upon and/or modified during the generation of pseudorandom bits (i.e., the working state).

7.4   The DRBG Mechanism Functions

The DRBG mechanism functions handle the DRBG's internal state. The DRBG mechanisms in this Recommendation have five separate functions:

1. The instantiate function acquires entropy input and may combine it with a nonce and a personalization string to create a seed from which the initial internal state is created.

2. The generate function generates pseudorandom bits upon request, using the current internal state, and generates a new internal state for the next request.

3. The reseed function acquires new entropy input and combines it with the current internal state and any additional input that is provided to create a new seed and a new internal state.

4. The uninstantiate function zeroizes (i.e., erases) the internal state.

5. The health test function determines that the DRBG mechanism continues to function correctly.

8. DRBG Mechanism Concepts and General Requirements

8.1 DRBG Mechanism Functions

A DRBG mechanism requires instantiate, uninstantiate, generate, and health testing functions. A DRBG mechanism may also include a reseed function. A DRBG **shall** be instantiated prior to the generation of output by the DRBG. These functions are specified in Section 9.

8.2 DRBG Instantiations

A DRBG may be used to obtain pseudorandom bits for different purposes (e.g., DSA private keys and AES keys) and may be separately instantiated for each purpose, thus effectively creating two DRBGs.

A DRBG is instantiated using a seed and may be reseeded; when reseeded, the seed **shall** be different than the seed used for instantiation. Each seed defines a *seed period* for the DRBG instantiation; an instantiation consists of one or more seed periods that begin when a new seed is acquired (see Figure 2).

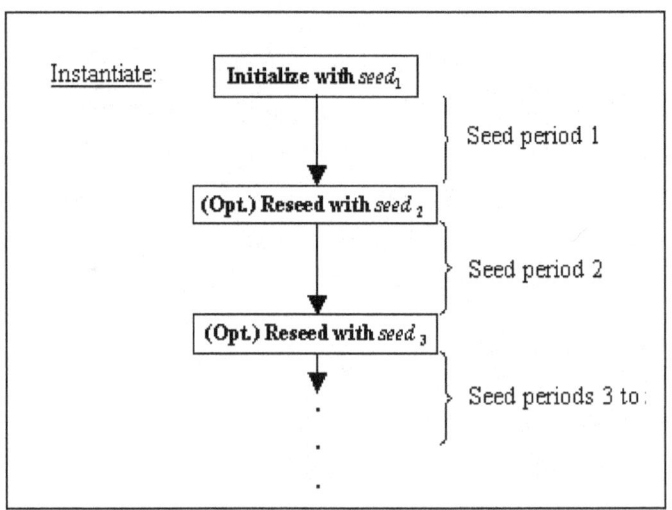

Figure 2: DRBG Instantiation

8.3 Internal States

During instantiation, an initial internal state is derived from the seed. The internal state for an instantiation includes:

1. The working state:

   a. One or more values that are derived from the seed and become part of the internal state; these values **shall** remain secret, and

   b. A count of the number of requests or blocks produced since the instantiation was seeded or reseeded.

2. Administrative information (e.g., security strength and prediction resistance flag).

The internal state **shall** be protected at least as well as the intended use of the pseudorandom output bits requested by the consuming application. A DRBG mechanism implementation may be designed to handle multiple instantiations. Each DRBG instantiation **shall** have its own internal state. The internal state for one DRBG instantiation **shall not** be used as the internal state for a different instantiation.

A DRBG transitions between internal states when the generator is requested to provide new pseudorandom bits. A DRBG may also be implemented to transition in response to internal or external events (e.g., system interrupts) or to transition continuously (e.g., whenever time is available to run the generator).

8.4   Security Strengths Supported by an Instantiation

The DRBG mechanisms specified in this Recommendation support four security strengths: 112, 128, 192 or 256 bits. The security strength for the instantiation is requested during DRBG instantiation, and the instantiate function obtains the appropriate amount of entropy for the requested security strength. Any security strength may be requested (up to a maximum of 256 bits), but the DRBG will only be instantiated to one of the four security strengths above, depending on the DRBG implementation. A requested security strength that is below the 112-bit security strength or is between two of the four security strengths will be instantiated to the next highest strength (e.g., a requested security strength of 80 bits will result in an instantiation at the 112-bit security strength).

The actual security strength supported by a given instantiation depends on the DRBG implementation and on the amount of entropy provided to the instantiate function. Note that the security strength actually supported by a particular instantiation could be less than the maximum security strength possible for that DRBG implementation (see Table 1). For example, a DRBG that is designed to support a maximum security strength of 256 bits could, instead, be instantiated to support only a 128-bit security strength if the additional security provided by the 256-bit security strength is not required (i.e., by requesting only 128 bits of entropy during instantiation, rather than 256 bits of entropy).

Table 1: Possible Instantiated Security Strengths

Maximum Designed Security Strength	112	128	192	256
Possible Instantiated Security Strengths	112	112, 128	112, 128, 192	112, 128, 192, 256

Following instantiation, requests can be made to the generate function of that instantiation for pseudorandom bits. For each generate request, the security strength to be provided for the bits is requested. Any security strength can be requested during a call to the generate function, up to the security strength of the instantiation, e.g., an instantiation could be instantiated at the 128-bit security strength, but a request for pseudorandom bits could indicate that a lesser security strength is actually required for the bits to be generated. The generate function checks that the requested security strength does not exceed the security strength for the instantiation. Assuming that the request is valid, the requested number of bits is returned.

When an instantiation is used for multiple purposes, the minimum entropy requirement for each purpose must be considered. The DRBG needs to be instantiated for the highest security strength required. For example, if one purpose requires a security strength of 112 bits, and another purpose requires a security strength of 256 bits, then the DRBG needs to be instantiated to support the 256-bit security strength.

8.5 DRBG Mechanism Boundaries

As a convenience, this Recommendation uses the notion of a "DRBG mechanism boundary" to explain the operations of a DRBG mechanism and its interaction with and relation to other processes; a DRBG mechanism boundary contains all DRBG mechanism functions and internal states required for a DRBG. Data enters a DRBG mechanism boundary via the DRBG's public interfaces, which are made available to consuming applications.

Within a DRBG mechanism boundary,

1. The DRBG internal state and the operation of the DRBG mechanism functions **shall** only be affected according to the DRBG mechanism specification.

2. The DRBG internal state **shall** exist solely within the DRBG mechanism boundary. The internal state **shall not** be accessible by non-DRBG functions or other instantiations of that or other DRBGs.

3. Information about secret parts of the DRBG internal state and intermediate values in computations involving these secret parts **shall not** affect any information that leaves the DRBG mechanism boundary, except as specified for the DRBG pseudorandom bit outputs.

Each DRBG mechanism includes one or more cryptographic primitives (e.g., a hash function or block cipher algortihm). Other applications may use the same cryptographic primitive, but the DRBG's internal state and the DRBG mechanism functions **shall not** be affected by these other applications.

A DRBG mechanism's functions may be contained within a single device, or may be distributed across multiple devices (see Figures 3 and 4). Figure 3 depicts a DRBG for which all functions are contained within the same device. Figure 4 provides an example of DRBG mechanism functions that are distributed across multiple devices. In this latter case, each device has a DRBG mechanism sub-boundary that contains the DRBG mechanism functions implemented on

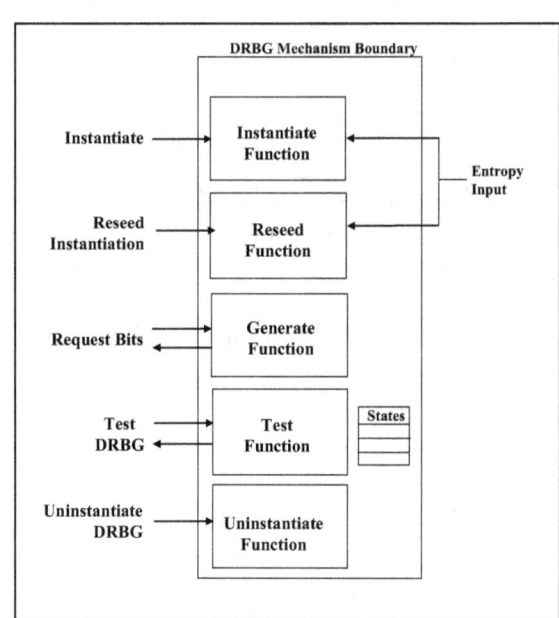

Figure 3: DRBG Mechanism Functions within a Single Device

that device. The boundary around the entire DRBG mechanism **shall** include the aggregation of sub-boundaries providing the DRBG mechanism functionality. The use of distributed DRBG mechanism functions may be convenient for restricted environments (e.g., smart card applications) in which the primary use of the DRBG does not require repeated use of the instantiate or reseed functions.

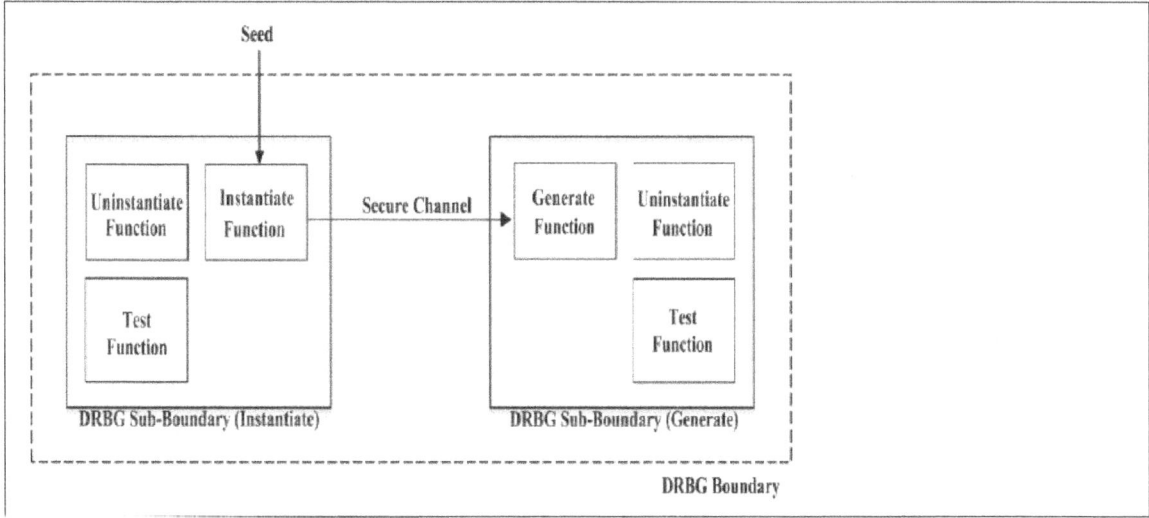

Figure 4: Distributed DRBG Mechanism Functions

Each DRBG mechanism boundary or sub-boundary **shall** contain a test function to test the "health" of other DRBG mechanism functions within that boundary. In addition, each boundary or sub-boundary **shall** contain an uninstantiate function in order to perform and/or react to health testing.

When DRBG mechanism functions are distributed, a secure channel **shall** be used to protect the confidentiality and integrity of the internal state or parts of the internal state that are transferred between the distributed DRBG mechanism sub-boundaries. The security provided by the secure channel **shall** be consistent with the security required by the consuming application.

8.6   Seeds

When a DRBG is used to generate pseudorandom bits, a seed **shall** be acquired prior to the generation of output bits by the DRBG. The seed is used to instantiate the DRBG and determine the initial internal state that is used when calling the DRBG to obtain the first output bits.

Reseeding is a means of restoring the secrecy of the output of the DRBG if a seed or the internal state becomes known. Periodic reseeding is a good way of addressing the threat of either the DRBG seed, entropy input or working state being compromised over time. In some implementations (e.g., smartcards), an adequate reseeding process may not be possible. In these cases, the best policy might be to replace the DRBG, obtaining a new seed in the process (e.g., obtain a new smart card).

The seed and its use by a DRBG mechanism **shall** be generated and handled as specified in the following subsections.

8.6.1   Seed Construction for Instantiation

Figure 5 depicts the seed construction process for instantiation. The seed material used to determine a seed for instantiation consists of entropy input, a nonce and an optional personalization string. Entropy input **shall** always be used in the construction of a seed; requirements for the entropy input are discussed in Section 8.6.3. Except for the case noted below, a nonce **shall** be used; requirements for the nonce are discussed in Section 8.6.7. A personalization string **should** also be used; requirements for the personalization string are discussed in Section 8.7.1.

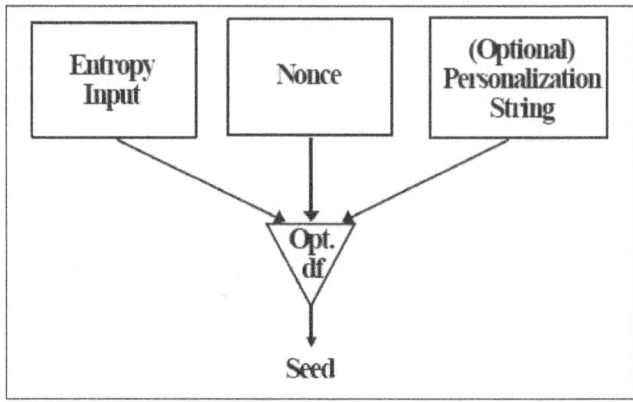

Figure 5: Seed Construction for Instantiation

Depending on the DRBG mechanism and the source of the entropy input, a derivation function may be required to derive a seed from the seed material. However, in certain circumstances, the DRBG mechanism based on block cipher algorithms (see Section 10.2) may be implemented without a derivation function. When implemented in this manner, a (separate) nonce (as shown in Figure 5) is not used. Note, however, that the personalization string could contain a nonce, if desired.

8.6.2   Seed Construction for Reseeding

Figure 6 depicts the seed construction process for reseeding an instantiation. The seed material for reseeding consists of a value that is carried in the internal state[3], new entropy input and, optionally, additional input. The internal state value and the entropy input are required; requirements for the entropy input are discussed in Section 8.6.3. Requirements for the additional input are discussed in Section 8.7.2. As in Section 8.6.1, a derivation function may be required for

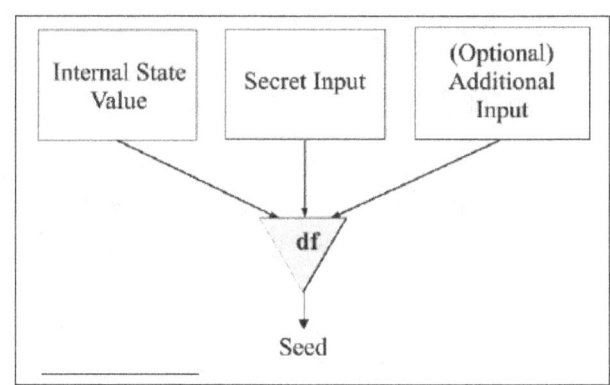

Figure 6: Seed Construction for Reseeding

---

[3] See each DRBG mechanism specification for the value that is used.

reseeding. See Section 8.6.1 for further guidance.

8.6.3   Entropy Requirements for the Entropy Input

The entropy input **shall** have entropy that is equal to or greater than the security strength of the instantiation. Additional entropy may be provided in the nonce or the optional personalization string during instantiation, or in the additional input during reseeding and generation, but this is not required. The use of more entropy than the minimum value will offer a security "cushion". This may be useful if the assessment of the entropy provided in the entropy input is incorrect. Having more entropy than the assessed amount is acceptable; having less entropy than the assessed amount could be fatal to security. The presence of more entropy than is required, especially during the instantiatiation, will provide a higher level of assurance than the minimum required entropy.

8.6.4   Seed Length

The minimum length of the seed depends on the DRBG mechanism and the security strength required by the consuming application. See Section 10.

8.6.5   Source of Entropy Input

The source of the entropy input (SEI) **shall** be either:

1. An **approved** entropy source,

2. An **approved** NRBG (note that an NRBG includes an entropy source), or

3. An **approved** DRBG, thus forming a chain of at least two DRBGs; the initial DRBG in the chain **shall** be seeded by an **approved** NRBG or an **approved** entropy source. A DRBG instantiation may seed or reseed another DRBG instantiation, but **shall not** reseed itself.

In cases 1 and 2, the SEI provides fresh entropy bits upon request. In case 3, the SEI can provide fresh entropy bits only if it has access to an entropy source or NRBG at the time of the request; otherwise, the entropy input provided by the SEI has a maximum security strength that is (at most) the security strength of the DRBG serving as the SEI. Further discussion about entropy and entropy sources is provided in [SP 800-90B]; further discussion on RBG construction and SEIs is provided in [SP 800-90C].

8.6.6   Entropy Input and Seed Privacy

The entropy input and the resulting seed **shall** be handled in a manner that is consistent with the security required for the data protected by the consuming application. For example, if the DRBG is used to generate keys, then the entropy inputs and seeds used to generate the keys **shall** (at a minimum) be protected as well as the keys.

8.6.7   Nonce

A nonce may be required in the construction of a seed during instantiation in order to provide a security cushion to block certain attacks. The nonce **shall** be either:

   a. A value with at least (1/2 *security_strength*) bits of entropy,

   b. A value that is expected to repeat no more often than a (1/2 *security_strength*)-bit random string would be expected to repeat.

For case a, the nonce may be acquired from the same source and at the same time as the entropy input. In this case, the seed could be considered to be constructed from an "extra strong" entropy input and the optional personalization string, where the entropy for the entropy input is equal to or greater than (3/2 *security_strength*) bits.

The nonce provides greater assurance that the DRBG provides *security_strength* bits of security to the consuming application. When a DRBG is instantiated many times without a nonce, a compromise may become more likely. In some consuming applications, a single DRBG compromise may reveal long-term secrets (e.g., a compromise of the DSA per-message secret may reveal the signing key).

8.6.8   Reseeding

Generating too many outputs from a seed (and other input information) may provide sufficient information for successfully predicting future outputs (see Section 8.8). Periodic reseeding will reduce security risks, reducing the likelihood of a compromise of the data that is protected by cryptographic mechanisms that use the DRBG.

Seeds **shall** have a finite seedlife (i.e., the number of blocks or outputs that are produced during a seed period); the maximum seedlife is dependent on the DRBG mechanism used. Reseeding is accomplished by 1) an explicit reseeding of the DRBG by the consuming application, or 2) by the generate function when prediction resistance is requested (see Section 8.8).

Reseeding of the DRBG **shall** be performed in accordance with the specification for the given DRBG mechanism. The DRBG reseed specifications within this Recommendation are designed to produce a new seed that is determined by both the old seed and newly obtained entropy input that will support the desired security strength.

An alternative to reseeding is to create an entirely new instantiation. However, reseeding is preferred over creating a new instantiation. If a DRBG instantiation was initially seeded with sufficient entropy, and the source of entropy input subsequently fails without being detected, then a new instantiation using the same (failed) source of entropy input would not have sufficient entropy to operate securely. However, if there is an undetected failure in the source of entropy input of an already properly seeded DRBG instantiation, the DRBG instantiation will still retain any previous entropy when the reseed operation fails to introduce new entropy.

8.6.9   Seed Use

The seed that is used to initialize one instantiation of a DRBG **shall not** be intentionally used to reseed the same instantiation or used as the seed for another DRBG instantiation. In addition, a DRBG **shall not** reseed itself. Note that a DRBG does not provide output until a seed is available, and the internal state has been initialized (see Section 10).

8.6.10   Entropy Input and Seed Separation

The seed used by a DRBG and the entropy input used to create that seed **shall not** intentionally be used for other purposes (e.g., domain parameter or prime number generation).

8.7   Other Input to the DRBG Mechanism

Other input may be provided during DRBG instantiation, pseudorandom bit generation and reseeding. This input may contain entropy, but this is not required. During instantiation, a personalization string may be provided and combined with entropy input and a nonce to derive a seed (see Section 8.6.1). When pseudorandom bits are requested and when reseeding is performed, additional input may be provided (see Section 8.7.2).

Depending on the method for acquiring the input, the exact value of the input may or may not be known to the user or consuming application. For example, the input could be derived directly from values entered by the user or consuming application, or the input could be derived from information introduced by the user or consuming application (e.g., from timing statistics based on key strokes), or the input could be the output of another RBG.

8.7.1   Personalization String

During instantiation, a personalization string **should** be used to derive the seed (see Section 8.6.1). The intent of a personalization string is to differentiate this DRBG instantiation from all other instantiations that might ever be created. The personalization string **should** be set to some bitstring that is as unique as possible, and may include secret information. Secret information **should not** be used in the personalization string if it requires a level of protection that is greater than the intended security strength of the DRBG instantiation. Good choices for the personalization string contents include:

- Device serial numbers,
- Public keys,
- User identification,
- Per-module or per-device values,
- Timestamps,
- Network addresses,

- Special key values for this specific DRBG instantiation,
- Application identifiers,
- Protocol version identifiers,
- Random numbers,
- Nonces,
- Seedfiles.

8.7.2   Additional Input

During each request for bits from a DRBG and during reseeding, the insertion of additional input is allowed. This input is optional, and the ability to enter additional input may or may not be included in an implementation.   Additional input may be either secret or publicly known; its value is arbitrary, although its length may be restricted, depending on the implementation and the DRBG mechanism. The use of additional input may be a means of providing more entropy for the DRBG internal state that will increase assurance that the entropy requirements are met. If the additional input is kept secret and has sufficient entropy, the input can provide more assurance when recovering from the compromise of the entropy input, the seed or one or more DRBG internal states.

8.8 Prediction Resistance and Backtracking Resistance

Figure 7 depicts the sequence of DRBG internal states that result from a given seed. Some subset of bits from each internal state are used to generate pseudorandom bits upon request by a user. The following discussions will use the figure to explain backtracking and prediction resistance.

Suppose that a compromise occurs at $State_x$, where $State_x$ contains both secret and non-secret information.

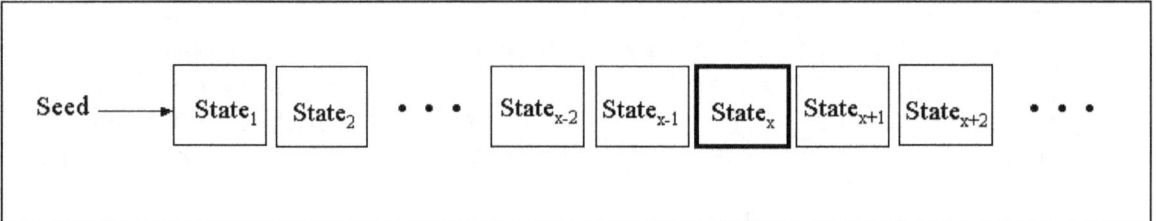

Figure 7: Sequence of DRBG States

Backtracking Resistance: Backtracking resistance is provided relative to time T if there is assurance that an adversary who has knowledge of the internal state of the DRBG at some time subsequent to time T would be unable to distinguish between observations of ideal random bitstrings and (previously unseen) bitstrings that were output by the DRBG prior to time T. This assumes that the adversary is incapable of performing the work required to negate the claimed security strength of the DRBG. Backtracking resistance means that a compromise of the DRBG internal state has no effect on the security of prior outputs. That is, an adversary who is given access to all of the prior output sequence cannot distinguish it from random output with less work than is associated with the security strength of the instantiation; if the adversary knows only part of the prior output, he cannot determine any bit of that prior output sequence that he has not already seen.

For example, suppose that an adversary knows $State_x$. Backtracking resistance means that:

    a.   The output bits from $State_1$ to $State_{x-1}$ cannot be distinguished from random output.

b. The prior internal state values themselves (*State*$_1$ to *State*$_{x-1}$) cannot be recovered, given knowledge of the secret information in *State*$_x$.

Backtracking resistance can be provided by ensuring that the DRBG generate algorithm is a one-way function. All DRBG mechanisms in this Recommendation have been designed to provide backtracking resistance.

Prediction Resistance: Prediction resistance means that a compromise of the DRBG internal state has no effect on the security of future DRBG outputs. That is, an adversary who is given access to all of the output sequence after the compromise cannot distinguish it from random output with less work than is associated with the security strength of the instantiation; if the adversary knows only part of the future output sequence, he cannot predict any bit of that future output sequence that he has not already known.

For example, suppose that an adversary knows *State*$_x$: Prediction resistance means that:

a. The output bits from *State*$_{x+1}$ and forward cannot be distinguished from an ideal random bitstring by the adversary.

b. The future internal state values themselves (*State*$_{x+1}$ and forward) cannot be predicted, given knowledge of *State*$_x$.

Prediction resistance is provided relative to time T if there is assurance that an adversary with knowledge of the state of the RBG at some time(s) prior to T (but incapable of performing work that matches the claimed *security strength* of the RBG) would be unable to distinguish between a observations of *ideal random bitstrings* and (previously unseen) bitstrings output by the RBG at or subsequent to time T. In particular, an RBG whose design allows the adversary to step forward from the initially compromised RBG state(s) to obtain knowledge of subsequent RBG states and the corresponding outputs (including the RBG state and output at time T) would <u>not</u> provide prediction resistance relative to time T.

Prediction resistance can be provided only by ensuring that a DRBG is effectively reseeded with fresh entropy between DRBG requests. That is, an amount of entropy that is sufficient to support the security strength of the DRBG being reseeded (i.e., an amount that is at least equal to the security strength) must be provided to the DRBG in a way that ensures that knowledge of the current DRBG internal state does not allow an adversary any useful knowledge about future DRBG internal states or outputs. Prediction resistance can be provided when the source of entropy input is or has access to an entropy source or an NRBG (see Section 8.6.5).

For example, suppose that an adversary knows internal *state*$_{x-2}$ (see Figure 7). If the adversary also knows the DRBG mechanism used, he then has enough information to compute *state*$_{x-1}$ and *state*$_x$. If prediction is then requested for the next bits that are to be output from the DRBG, new entropy bits will be inserted into the DRBG instantiation that will create a "barrier" between *state*$_x$ and *state*$_{x+1}$, i.e., the adversary will not be able compute *state*$_{x+1}$, simply by knowing *state*$_x$ – the work required will be greatly increased by the entropy inserted during the prediction request.

9   DRBG Mechanism Functions

Except for the health test function, which is discussed in Section 11.3, the functions of the DRBG mechanisms in this Recommendation are specified as an algorithm and an "envelope" of pseudocode around that algorithm. The pseudocode in the envelopes (provided in this section) checks the input parameters, obtains input not provided via the input parameters, accesses the appropriate DRBG algorithm and handles the internal state. A function need not be implemented using such envelopes, but the function **shall** have equivalent functionality.

During instantiation and reseeding (see Sections 9.1 and 9.2), entropy input is acquired for constructing a seed as discussed in Sections 8.6.1 and 8.6.2. In the specifications of this Recommendation, a **Get_entropy_input** pseudo-function is used for this purpose. The entropy input **shall not** be provided by a consuming application as an input parameter in an instantiate or reseed request. The **Get_entropy_input** function is fully specified in [SP 800-90C] for various RBG constructions; however, in general, the function has the following meaning:

> **Get_entropy_input**: A function that is used to obtain entropy input. The function call is:
>
> > ($status$, $entropy\_input$) = **Get_entropy_input** ($min\_entropy$, min_length, max_length, $prediction\_resistance\_request$),

which requests a string of bits ($entropy\_input$) with at least $min\_entropy$ bits of entropy. The length for the string **shall** be equal to or greater than min_length bits, and less than or equal to max_length bits. The $prediction\_resistance\_request$ parameter indicates whether or not prediction resistance is to be provided during the request (i.e., whether fresh entropy is required[4]). A $status$ code is also returned from the function.

Note that an implementation may choose to define this functionality differently by omitting some of the parameters; for example, for many of the DRBG mechanisms, $min\_length = min\_entropy$ for the **Get_entropy_input** function, in which case, the second parameter could be omitted.

In the pseudocode in this section, two classes of error codes are returned: **ERROR_FLAG** and **CATASTROPHIC_ERROR_FLAG**. These error codes are discussed in Section 11.3.6.

Comments are often included in the pseudocode in this Recommendation. A comment placed on a line that includes pseudocode applies to that line; a comment placed on a line containing no pseudocode applies to one or more lines of pseudocode immediately below that comment.

---

[4] Entropy input may be obtained from an entropy source or an NRBG, both of which provide fresh entropy. Entropy input could also be obtained from a DRBG that may or may not have access to an entropy source or NRBG. The request for prediction resistance rules out the use of a DRBG that does not have access to either an entropy source or NRBG.

9.1 Instantiating a DRBG

A DRBG **shall** be instantiated prior to the generation of pseudorandom bits. The instantiate function:

1. Checks the validity of the input parameters,

2. Determines the security strength for the DRBG instantiation,

3. Determines any DRBG mechanism-specific parameters (e.g., elliptic curve domain parameters),

4. Obtains entropy input with entropy sufficient to support the security strength,

5. Obtains the nonce (if required),

6. Determines the initial internal state using the instantiate algorithm,

7. If an implemention supports multiple simultaneous instantiations of the same DRBG, a *state_handle* for the internal state is returned to the consuming application (see below).

Let *working_state* be the working state for the particular DRBG mechanism (e.g., **HMAC_DRBG**), and let *min_length*, *max_ length*, and *highest_supported_security_strength* be defined for each DRBG mechanism (see Section 10). Let **Instantiate_algorithm** be a call to the appropriate instantiate algorithm for the DRBG mechanism (see Section 10).

The following or an equivalent process **shall** be used to instantiate a DRBG.

Instantiate_function (*requested_instantiation_security_strength, prediction_resistance_flag, personalization_string*):

1. *requested_instantiation_security_strength*: A requested security strength for the instantiation. Implementations that support only one security strength do not require this parameter; however, any consuming application using that implementation must be aware of the security strength that is supported.

2. *prediction_resistance_flag*: Indicates whether or not prediction resistance may be required by the consuming application during one or more requests for pseudorandom bits. Implementations that always provide or do not support prediction resistance do not require this parameter, since the intent is implicitly known. However, the user of a consuming application must determine whether or not prediction resistance may be required by the consuming application before electing to use such an implementation. If the *prediction_resistance_flag* is not needed (i.e., because prediction resistance is always performed or is not supported), then the *prediction_resistance_flag* input parameter and instantiate process step 2 are omitted, and the *prediction_resistance_flag* is omitted from the internal state in step 11 of the instantiate process. In addition, step 6 can be modified to not perform a check for the *prediction_resistance_flag* when the flag is not used in an implementation; in this case, the **Get_entropy_input** call need not include the *prediction_resistance_request* parameter.

3. *personalization_string*: An optional input that provides personalization information (see Sections 8.6.1 and 8.7.1). The maximum length of the personalization string (*max_personalization_string_length*) is implementation dependent, but **shall** be less than or equal to the maximum length specified for the given DRBG mechanism (see Section 10). If the input of a personalization string is not supported, then the *personalization_string* input parameter and step 3 of the instantiate process are omitted, and instantiate process step 9 is modified to omit the personalization string.

Required information not provided by the consuming application during instantiation (This information **shall not** be provided by the consuming application as an input parameter during the instantiate request):

1. *entropy_input*: Input bits containing entropy. The maximum length of the *entropy_input* is implementation dependent, but **shall** be less than or equal to the specified maximum length for the selected DRBG mechanism (see Section 10).

2. *nonce*: A nonce as specified in Section 8.6.7. Note that if a random value is used as the nonce, the *entropy_input* and *nonce* could be acquired using a single **Get_entropy_input** call (see step 6 of the instantiate process); in this case, the first parameter of the **Get_entropy_input** call is adjusted to include the entropy for the *nonce* (i.e., the *security_strength* is increased by at least ½ *security_strength*, and *min-length* is increased to accommodate the length of the nonce), instantiate process step 8 is omitted, and the *nonce* is omitted from the parameter list in instantiate process step 9.

   Note that in some cases, a nonce will not be used by a DRBG mechanism; in this case, step 8 is omitted, and the *nonce* is omitted from the parameter list in instantiate process step 9.

Output to a consuming application after instantiation:

1. *status*: The status returned from the instantiate function. The *status* will indicate **SUCCESS** or an **ERROR** (i.e., either **ERROR_FLAG** or **CATASTROPHIC_ERROR_FLAG**). If an **ERROR** is indicated, either no *state_handle* or an invalid *state_handle* **shall** be returned. A consuming application **should** check the *status* to determine that the DRBG has been correctly instantiated.

2. *state_handle*: Used to identify the internal state for this instantiation in subsequent calls to the generate, reseed, uninstantiate and test functions.

   If a state handle is not required for an implementation because the implementation does not support multiple simultaneous instantiations, a *state_handle* need not be returned. In this case, instantiate process step 10 is omitted, process step 11 is revised to save the only internal state, and process step 12 is altered to omit the *state_handle*.

Information retained within the DRBG mechanism boundary after instantiation:

The internal state for the DRBG, including the *working_state* and administrative information (see Sections 8.3 and 10 for definitions of the *working_state* and administrative information).

Instantiate Process:

Comment: Check the validity of the input parameters.

1. If *requested_instantiation_security_strength* > *highest_supported_security_strength*, then return an **ERROR_FLAG**.

2. If *prediction_resistance_flag* is set, and prediction resistance is not supported, then return an **ERROR_FLAG**.

3. If the length of the *personalization_string* > *max_personalization_string_length*, return an **ERROR_FLAG**.

4. Set *security_strength* to the lowest security strength greater than or equal to *requested_instantiation_security_strength* from the set {112, 128, 192, 256}.

5. Null step.

Comment: This is intended to replace a step from the previous version without changing the step numbers.[5]

Comment: Obtain the entropy input.

6. (*status*, *entropy_input*) = **Get_entropy_input** (*security_strength*, *min_length*, *max_length*, *prediction_resistance_request*).

7. If an **ERROR** is returned in step 6, return a **CATASTROPHIC_ERROR_FLAG**.

8. Obtain a *nonce*.

Comment: This step **shall** include any appropriate checks on the acceptability of the *nonce*. See Section 8.6.7.

Comment: Call the appropriate instantiate algorithm in Section 10 to obtain values for the initial *working_state*.

9. *initial_working_state* = **Instantiate_algorithm** (*entropy_input*, *nonce*, *personalization_string*, *security_strength*).

10. Get a *state_handle* for a currently empty internal state. If an empty internal state cannot be found, return an **ERROR_FLAG**.

---

[5] This step was originally used for selecting a curve when the Dual_EC_DRBG was implemented. However, this task is more appropriately done by the Dual_EC_DRBG's Instantiate_algorithm, after which the parameters for the internal state are returned to the Instantiate_function. This change is not intended to invalidate implementations, but to provide an equivalent process.

11. Set the internal state for the new instantiation (e.g., as indicated by *state_handle*) to the initial values for the internal state (i.e., set the *working_state* to the values returned as *initial_working_state* in step 9 and any other values required for the *working_state* (see Section 10), and set the administrative information to the appropriate values (e.g., the values of *security_strength* and the *prediction_resistance_flag*).

12. Return **SUCCESS** and *state_handle*.

9.2   Reseeding a DRBG Instantiation

The reseeding of an instantiation is not required, but is recommended whenever a comsuming application and implementation are able to perform this process. Reseeding will insert additional entropy input into the generation of pseudorandom bits. Reseeding may be:

- explicitly requested by a consuming application,

- performed when prediction resistance is requested by a consuming application,

- triggered by the generate function when a predetermined number of pseudorandom outputs have been produced or a predetermined number of generate requests have been made (i.e., at the end of the seedlife), or

- triggered by external events (e.g., whenever sufficient entropy is available).

If a reseed capability is not supported, a new DRBG instantiation may be created (see Section 9.1).

The reseed function:

1. Checks the validity of the input parameters,

2. Obtains entropy input from a source that supports the security strength of the DRBG, and

3. Using the reseed algorithm, combines the current working state with the new entropy input and any additional input to determine the new working state.

Let *working_state* be the working state for the particular DRBG instantiation (e.g., **HMAC_DRBG**) , let *min_length* and *max_ length* be defined for each DRBG mechanism, and let **Reseed_algorithm** be a call to the appropriate reseed algorithm for the DRBG mechanism (see Section 10).

The following or an equivalent process **shall** be used to reseed the DRBG instantiation.

Reseed_function (*state_handle, prediction_resistance_request, additional_input*):

1) *state_handle*: A pointer or index that indicates the internal state to be reseeded. If a state handle is not used by an implementation because the implemention does not support multiple simultaneous instantiations, a *state_handle* is not provided as input. Since there is only a single internal state in this case, reseed process step 1

obtains the contents of the internal state, and process step 6 replaces the *working_state* of this internal state.

2) *prediction_resistance_request*: Indicates whether or not prediction resistance is to be provided during the request (i.e., whether or not fresh entropy bits are required)[6]. DRBGs that are implemented to always support prediction resistance or to never support prediction resistance do not require this parameter. However, when prediction resistance is not supported, the user of a consuming application must determine whether or not prediction resistance may be required by the application before electing to use such a DRBG implementation.

    If prediction resistance is not supported, then the *prediction_resistance_request* input parameter and step 2 of the reseed process is omitted, and reseed process step 4 is modified to omit the *prediction_resistance_request* parameter.

    If prediction resistance is always performed, then the *prediction_resistance_request* input parameter and reseed process step 2 may be omitted, and reseed process step 4 is replaced by:

    (*status*, *entropy_input*) = **Get_entropy_input** (*security_strength*, *min_length*, *max_length*)

3) *additional_input*: An optional input. The maximum length of the *additional_input* (*max_additional_input_length*) is implementation dependent, but **shall** be less than or equal to the maximum value specified for the given DRBG mechanism (see Section 10). If the input by a consuming application of *additional_input* is not supported, then the input parameter and step 2 of the reseed process are omitted, and step 5 of the reseed process is modified to remove the *additional_input* from the parameter list.

Required information not provided by the consuming application during reseeding (This information **shall not** be provided by the consuming application as an input parameter during the reseed request):

1. *entropy_input*: Input bits containing entropy. This input **shall not** be provided by the DRBG instantiation being reseeded. The maximum length of the *entropy_input* is implementation dependent, but **shall** be less than or equal to the specified maximum length for the selected DRBG mechanism (see Section 10).

2. Internal state values required by the DRBG for the *working_state* and administrative information, as appropriate.

Output to a consuming application after reseeding:

1. *status*: The status returned from the function. The *status* will indicate **SUCCESS** or an **ERROR**.

---

[6] A DRBG may be reseeded by an entropy source or an NRBG, both of which provide fresh entropy. A DRBG could also be reseeded by a DRBG that may or may not have access to an entropy source or NRBG. The request for prediction resistance during reseeding rules out the use of a DRBG that does not have access to either an entropy source or NRBG. See [SP 800-90C] for further discussion.

Information retained within the DRBG mechanism boundary after reseeding:

Replaced internal state values (i.e., the *working_state*).

Reseed Process:

> Comment: Get the current internal state and check the input parameters.

1. Using *state_handle*, obtain the current internal state. If *state_handle* indicates an invalid or unused internal state, return an **ERROR_FLAG**.

2. If *prediction_resistance_request* is set, and *prediction_resistance_flag* is not set, then return an **ERROR_FLAG**.

3. If the length of the *additional_input* > *max_additional_input_length*, return an **ERROR_FLAG**.

> Comment: Obtain the entropy input.

4. (*status, entropy_input*) = **Get_entropy_input** (*security_strength, min_length, max_length, prediction_resistance_request*).

5. If an **ERROR** is returned in step 4, return a **CATASTROPHIC_ERROR_FLAG**.

> Comment: Get the new *working_state* using the appropriate reseed algorithm in Section 10.

6. *new_working_state* = **Reseed_algorithm** (*working_state, entropy_input, additional_input*).

7. Replace the *working_state* in the internal state for the DRBG instantiation (e.g., as indicated by *state_handle*) with the values of *new_working_state* obtained in step 6.

8. Return **SUCCESS**.

9.3   Generating Pseudorandom Bits Using a DRBG

This function is used to generate pseudorandom bits after instantiation or reseeding. The generate function:

1. Checks the validity of the input parameters.

2. Calls the reseed function to obtain sufficient entropy if the instantiation needs additional entropy because the end of the seedlife has been reached or prediction resistance is required; see Sections 9.3.2 and 9.3.3 for more information on reseeding at the end of the seedlife and on handling prediction resistance requests.

3. Generates the requested pseudorandom bits using the generate algorithm.

4. Updates the working state.

5. Returns the requested pseudorandom bits to the consuming application.

9.3.1   The Generate Function

Let *outlen* be the length of the output block of the cryptographic primitive (see Section 10). Let **Generate_algorithm** be a call to the appropriate generate algorithm for the DRBG mechanism (see Section 10), and let **Reseed_function** be a call to the reseed function in Section 9.2.

The following or an equivalent process **shall** be used to generate pseudorandom bits.

Generate_function (*state_handle, requested_number_of_bits,*
     requested_security_strength, *prediction_resistance_request, additional_input*):

1. *state_handle*: A pointer or index that indicates the internal state to be used. If a state handle is not used by an implementation because the implemention does not support multiple simultaneous instantiations, a *state_handle* is not provided as input. The *state_handle* is then omitted from the input parameter list in process step 7.1, generate process steps 1 and 7.3 are used to obtain the contents of the internal state, and process step 10 replaces the *working_state* of this internal state.

2. *requested_number_of_bits*: The number of pseudorandom bits to be returned from the generate function. The *max_number_of_bits_per_request* is implementation dependent, but **shall** be less than or equal to the value provided in Section 10 for a specific DRBG mechanism.

3. *requested_security_strength*: The security strength to be associated with the requested pseudorandom bits. DRBG implementations that support only one security strength do not require this parameter; however, any consuming application using that DRBG implementation must be aware of the supported security strength.

4. *prediction_resistance_request*: Indicates whether or not prediction resistance is to be provided during the request. DRBGs that are implemented to always provide prediction resistance or that do not support prediction resistance do not require this parameter. However, when prediction resistance is not supported, the user of a consuming application must determine whether or not prediction resistance may be required by the application before electing to use such a DRBG implementation.

    If prediction resistance is not supported, then the *prediction_resistance_request* input parameter and steps 5 and 9.2 of the generate process are omitted, and generate process steps 7 and 7.1 are modified to omit the check for the *prediction_resistance_request* term.

    If prediction resistance is always performed, then the *prediction_resistance_request* input parameter and generate process steps 5 and 9.2 may be omitted, and generate process steps 7 and 8 may be replaced by:

    status = **Reseed_function** (*state_handle, additional_input*).

    If *status* indicates an **ERROR**, then return *status*.

Using *state_handle*, obtain the new internal state.

(*status, pseudorandom_bits, new_working_state*) = **Generate_algorithm** (*working_state, requested_number_of_bits*).

Note that if the input of *additional_input* is not supported, then the *additional_input* parameter in the **Reseed_function** call above may be omitted.

5. *additional_input*: An optional input. The maximum length of the *additional_input* (*max_additional_input_length*) is implementation dependent, but **shall** be less than or equal to the specified maximum length for the selected DRBG mechanism (see Section 10). If the input of *additional_input* is not supported, then the input parameter, generate process steps 4 and 7.4, and the *additional_input* input parameter in generate process steps 7.1 and 8 are omitted.

Required information not provided by the consuming application during generation:

1. Internal state values required for the *working_state* and administrative information, as appropriate.

Output to a consuming application after generation:

1. *status*: The status returned from the generate function. The *status* will indicate **SUCCESS** or an **ERROR**.

2. *pseudorandom_bits*: The pseudorandom bits that were requested.

Information retained within the DRBG mechanism boundary after generation:

Replaced internal state values (i.e., the new *working_state*).

Generate Process:

> Comment: Get the internal state and check the input parameters.

1. Using *state_handle*, obtain the current internal state for the instantiation. If *state_handle* indicates an invalid or unused internal state, then return an **ERROR_FLAG**.

2. If *requested_number_of_bits* > *max_number_of_bits_per_request*, then return an **ERROR_FLAG**.

3. If *requested_security_strength* > the *security_strength* indicated in the internal state, then return an **ERROR_FLAG**.

4. If the length of the *additional_input* > *max_additional_input_length*, then return an **ERROR_FLAG**.

5. If *prediction_resistance_request* is set, and *prediction_resistance_flag* is not set, then return an **ERROR_FLAG**.

6. Clear the *reseed_required_flag*.   Comment: See Section 9.3.2 for discussion.

> Comment: Reseed if necessary (see Section 9.2).

7. If *reseed_required_flag* is set, or if *prediction_resistance_request* is set, then

    7.1   *status* = **Reseed_function** (*state_handle*, *prediction_resistance_request*, *additional_input*).

    7.2   If *status* indicates an **ERROR**, then return *status*.

    7.3   Using *state_handle*, obtain the new internal state.

    7.4   *additional_input* = the *Null* string.

    7.5   Clear the *reseed_required_flag*.

> Comment: Request the generation of *pseudorandom_bits* using the appropriate generate algorithm in Section 10.

8. (*status, pseudorandom_bits, new_working_state*) = **Generate_algorithm** (*working_state, requested_number_of_bits, additional_input*).

9. If *status* indicates that a reseed is required before the requested bits can be generated, then

    9.1   Set the *reseed_required_flag*.

    9.2   If the *prediction_resistance_flag* is set, then set the *prediction_resistance request* indication.

    9.3   Go to step 7.

10. Replace the old *working_state* in the internal state of the DRBG instantiation (e.g., as indicated by *state_handle*) with the values of *new_working_state*.

11. Return **SUCCESS** and *pseudorandom_bits*.

Implementation notes:

If a reseed capability is not supported, or a reseed is not desired, then generate process steps 6 and 7 are removed; and generate process step 9 is replaced by:

9. If *status* indicates that a reseed is required before the requested bits can be generated, then

    9.1 *status* = **Uninstantiate_function** (*state_handle*).

    9.2 Return an indication that the DRBG instantiation can no longer be used.

9.3.2   Reseeding at the End of the Seedlife

When pseudorandom bits are requested by a consuming application, the generate function checks whether or not a reseed is required by comparing the counter within the internal state (see Section 8.3) against a predetermined reseed interval for the DRBG implementation. This is specified in the generate process (see Section 9.3.1) as follows:

a.   Step 6 clears the *reseed_required_flag*.

b. Step 7 checks the value of the *reseed_required_flag*. At this time, the *reseed_required_flag* is clear, so step 7 is skipped unless prediction resistance was requested by the consuming application. For the purposes of this explanation, assume that prediction resistance was not requested.

c. Step 8 calls the **Generate_algorithm**, which checks whether a reseed is required. If it is required, an appropriate *status* is returned.

d. Step 9 checks the *status* returned by the **Generate_algorithm**. If the *status* does not indicate that a reseed is required, the generate process continues with step 10.

e. However, if the status indicates that a reseed is required, then the *reseed_required_flag* is set, the *prediction_resistance_request* indicator is set if the instantiation is capable of performing prediction resistance, and processing continues by going back to step 7 (see steps 9.1 – 9.3). This is intended to obtain fresh entropy for reseeding at the end of the reseed interval whenever access to fresh entropy is available (see the concept of Live Entropy sources in [SP 800-90C]).

f. The substeps in step 7 are executed. The reseed function is called; any *additional_input* provided by the consuming application in the generate request is used during reseeding. The new values of the internal state are acquired, any *additional_input* provided by the consuming application in the generate request is replaced by a *Null* string, and the *reseed_required_flag* is cleared.

g. The generate algorithm is called (again) in step 8, the check of the returned *status* is made in step 9, and (presumably) step 10 is then executed.

9.3.3   Handling Prediction Resistance Requests

When pseudorandom bits are requested by a consuming application with prediction resistance, the generate function specified in Section 9.3.1 checks that the instantiation allows prediction resistance requests (see step 5 of the generate process); clears the *reseed_required_flag* (even though the flag won't be used in this case); executes the substeps of generate process step 7, resulting in a reseed, a new internal state for the instantiation, and setting the additional input to a *Null* value; obtains pseudorandom bits (see generate process step 8); passes through generate process step 9, since another reseed will not be required; and continues with generate process step 10.

9.4 Removing a DRBG Instantiation

The internal state for an instantiation may need to be "released" by erasing (i.e., zeroizing) the contents of the internal state. The uninstantiate function:

1. Checks the input parameter for validity, and

2. Empties the internal state.

The following or an equivalent process **shall** be used to remove (i.e., uninstantiate) a DRBG instantiation:

Uninstantiate_function (*state_handle*) :

1. *state_handle*: A pointer or index that indicates the internal state to be "released". If a state handle is not used by an implementation because the implemention does not support multiple simultaneous instantiations, a *state_handle* is not provided as input. In this case, process step 1 is omitted, and process step 2 erases the internal state.

Output to a consuming application after uninstantiation:

1. *status*: The status returned from the function. The status will indicate **SUCCESS** or **ERROR_FLAG**.

Information retained within the DRBG mechanism boundary after uninstantiation:

An empty internal state.

Uninstantiate Process:

1. If *state_handle* indicates an invalid state, then return an **ERROR_FLAG**.

2. Erase the contents of the internal state indicated by *state_handle*.

3. Return **SUCCESS**.

10 DRBG Algorithm Specifications

Several DRBG mechanisms are specified in this Recommendation. The selection of a DRBG mechanism depends on several factors, including the security strength to be supported and what cryptographic primitives are available. An analysis of the consuming application's requirements for random numbers **should** be conducted in order to select an appropriate DRBG mechanism. Conversion specifications required for the DRBG mechanism implementations (e.g., between integers and bitstrings) are provided in Appendix B. Pseudocode examples for each DRBG mechanism are provided in Appendix D. A detailed discussion on DRBG mechanism selection is provided in Appendix E.

10.1 DRBG Mechanisms Based on Hash Functions

A DRBG mechanism may be based on a hash function that is non-invertible or one-way. The hash-based DRBG mechanisms specified in this Recommendation have been designed to use any **approved** hash function and may be used by consuming applications requiring various security strengths, providing that the appropriate hash function is used and sufficient entropy is obtained for the seed.

The following are provided as DRBG mechanisms based on hash functions:

1. The **Hash_DRBG** specified in Section 10.1.1.

2. The **HMAC_DRBG** specified in Section 10.1.2.

The maximum security strength that can be supported by each DRBG based on a hash function is the security strength of the hash function used; the security strengths for the hash functions when used for random number generation are provided in [SP 800-57]. However, this Recommendation supports only four security strengths: 112, 128, 192, and 256 bits. Table 2 specifies the values that **shall** be used for the function envelopes and DRBG algorithm for each **approved** hash function.

Table 2: Definitions for Hash-Based DRBG Mechanisms

	SHA-1	SHA-224 and SHA-512/224	SHA-256 and SHA-512/256	SHA-384	SHA-512
Supported security strengths	See [SP 800-57]				
highest_supported_security_strength	See [SP 800-57]				
Output Block Length (*outlen*)	160	224	256	384	512
Required minimum entropy for instantiate and reseed	*security_strength*				
Minimum entropy input length (*min_length*)	*security_strength*				

	SHA-1	SHA-224 and SHA-512/224	SHA-256 and SHA-512/256	SHA-384	SHA-512
Maximum entropy input length (*max_ length*)	$\leq 2^{35}$ bits				
Seed length (*seedlen*) for Hash_DRBG	440	440	440	888	888
Maximum personalization string length (*max_personalization_string_length*)	$\leq 2^{35}$ bits				
Maximum additional_input length (*max_additional_input_length*)	$\leq 2^{35}$ bits				
max_number_of_bits_per_request	$\leq 2^{19}$ bits				
Number of requests between reseeds (*reseed_interval*)	$\leq 2^{48}$				

Note that since SHA-224 is based on SHA-256, there is no efficiency benefit when using SHA-224, rather than SHA-256. Also note that since SHA-384, SHA-512/224 and SHA-512/256 are based on SHA-512, there is no efficiency benefit for using these three SHA mechanisms, rather than using SHA-512. However, efficiency is just one factor to consider when selecting the appropriate hash function to use as part of a DRBG mechanism.

10.1.1   Hash_DRBG

Figure 8 presents the normal operation of the **Hash_DRBG** generate algorithm. The **Hash_DRBG** requires the use of a hash function during the instantiate, reseed and generate functions; the same hash function **shall** be used throughout a **Hash_DRBG** instantiation. **Hash_DRBG** uses the derivation function specified in Section 10.4.1 during instantiation and reseeding. The hash function to be used **shall** meet or exceed the desired security strength of the consuming application.

10.1.1.1   Hash_DRBG Internal State

The *internal_state* for **Hash_DRBG** consists of:

1. The *working state*:

   a. A value (*V*) of *seedlen* bits that is updated during each call to the DRBG.

   b. A constant *C* of *seedlen* bits that depends on the *seed*.

   c. A counter (*reseed_counter*) that indicates the number of requests for pseudorandom bits since new *entropy_input* was obtained during instantiation or reseeding.

2. Administrative information:

    a. The *security_strength* of the DRBG instantiation.

    b. A *prediction_resistance_flag* that indicates whether or not a prediction resistance capability is available for the DRBG instantiation.

The values of *V* and *C* are the critical values of the internal state upon which the security of this DRBG mechanism depends (i.e., *V* and *C* are the "secret values" of the internal state).

10.1.1.2 Instantiation of Hash_DRBG

Notes for the instantiate function specified in Section 9.1:

The instantiation of **Hash_DRBG** requires a call to the **Instantiate_function** specified in Section 9.1. Process step 9 of that function calls the instantiate algorithm in this section.

The values of *highest_supported_security_strength* and *min_length* are provided in Table 2 of Section 10.1. The contents of the internal state are provided in Section 10.1.1.1.

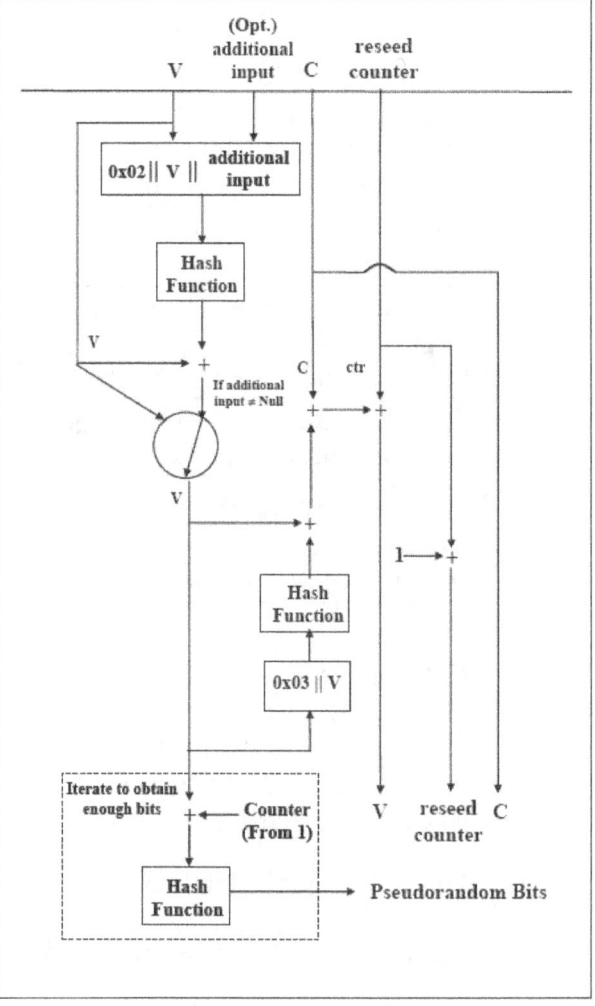

Figure 8: Hash_DRBG

The instantiate algorithm:

Let **Hash_df** be the hash derivation function specified in Section 10.4.1 using the selected hash function. The output block length (*outlen*), seed length (*seedlen*) and appropriate *security_strengths* for the implemented hash function are provided in Table 2 of Section 10.1.

The following process or its equivalent **shall** be used as the instantiate algorithm for this DRBG mechanism (see step 9 of the instantiate process in Section 9.1).

Hash_DRBG_Instantiate_algorithm (*entropy_input, nonce, personalization_string, security_strength*):

1. *entropy_input*: The string of bits obtained from the source of entropy input.

2. *nonce*: A string of bits as specified in Section 8.6.7.

3. *personalization_string*: The personalization string received from the consuming application. Note that the length of the *personalization_string* may be zero.

4. *security_strength*: The security strength for the instantiation. This parameter is optional for **Hash_DRBG**, since it is not used.

Output:

1. *initial_working_state*: The inital values for *V*, *C*, and *reseed_counter* (see Section 10.1.1.1).

Hash_DRBG Instantiate Process:

1. *seed_material = entropy_input || nonce || personalization_string*.

2. *seed* = **Hash_df** (*seed_material, seedlen*).

3. *V = seed*.

4. *C* = **Hash_df** ((0x00 || *V*), *seedlen*).   Comment: Preceed *V* with a byte of zeros.

5. *reseed_counter* = 1.

6. **Return** *V*, *C*, and *reseed_counter* as the *initial_working_state*.

10.1.1.3   Reseeding a Hash_DRBG Instantiation

Notes for the reseed function specified in Section 9.2:

The reseeding of a **Hash_DRBG** instantiation requires a call to the **Reseed_function**. Process step 6 of that function calls the reseed algorithm specified in this section. The values for *min_length* are provided in Table 2 of Section 10.1.

The reseed algorithm:

Let **Hash_df** be the hash derivation function specified in Section 10.4.1 using the selected hash function. The value for *seedlen* is provided in Table 2 of Section 10.1.

The following process or its equivalent **shall** be used as the reseed algorithm for this DRBG mechanism (see step 6 of the reseed process in Section 9.2):

Hash_DRBG_Reseed_algorithm (*working_state, entropy_input, additional_input*):

1. *working_state*: The current values for *V*, *C*, and *reseed_counter* (see Section 10.1.1.1).

2. *entropy_input*: The string of bits obtained from the source of entropy input.

3. *additional_input*: The additional input string received from the consuming application. Note that the length of the *additional_input* string may be zero.

Output:

1. *new_working_state*: The new values for *V*, *C*, and *reseed counter*.

Hash_DRBG Reseed Process:

1. *seed_material* = 0x01 ∥ *V* ∥ *entropy_input* ∥ *additional_input*.

2. *seed* = **Hash_df** (*seed_material, seedlen*).

3. *V* = *seed*.

4. *C* = **Hash_df** ((0x00 ∥ *V*), *seedlen*).   Comment: Preceed with a byte of all zeros.

5. *reseed_counter* = 1.

6. Return *V*, *C*, and *reseed_counter* for the *new_working_state*.

10.1.1.4   Generating Pseudorandom Bits Using Hash_DRBG

Notes for the generate function specified in Section 9.3:

The generation of pseudorandom bits using a **Hash_DRBG** instantiation requires a call to the generate function. Process step 8 of that function calls the generate algorithm specified in this section. The values for *max_number_of_bits_per_request* and *outlen* are provided in Table 2 of Section 10.1.

The generate algorithm:

Let **Hash** be the selected hash function. The seed length (*seedlen*) and the maximum interval between reseeding (*reseed_interval*) are provided in Table 2 of Section 10.1. Note that for this DRBG mechanism, the reseed counter is used to update the value of *V*, as well as to count the number of generation requests.

The following process or its equivalent **shall** be used as the generate algorithm for this DRBG mechanism (see step 8 of the generate process in Section 9.3):

Hash_DRBG_Generate_algorithm (*working_state, requested_number_of_bits, additional_input*):

1. *working_state*: The current values for *V*, *C*, and *reseed_counter* (see Section 10.1.1.1).

2. *requested_number_of_bits*: The number of pseudorandom bits to be returned to the generate function.

3. *additional_input*: The additional input string received from the consuming application. Note that the length of the *additional_input* string may be zero.

Output:

1. *status*: The status returned from the function. The *status* will indicate **SUCCESS,** or indicate that a reseed is required before the requested pseudorandom bits can be generated.

2. *returned_bits*: The pseudorandom bits to be returned to the generate function.

3. *new_working_state*: The new values for *V*, *C*, and *reseed_counter*.

Hash_DRBG Generate Process:

1. If *reseed_counter* > *reseed_interval*, then return an indication that a reseed is required.

2. If (*additional_input* ≠ *Null*), then do

   2.1 $w =$ **Hash** (0x02 $\|$ V $\|$ *additional_input*).

   2.2 $V = (V + w) \bmod 2^{seedlen}$.

3. (*returned_bits*) = **Hashgen** (*requested_number_of_bits*, V).

4. $H =$ **Hash** (0x03 $\|$ V).

5. $V = (V + H + C + reseed\_counter) \bmod 2^{seedlen}$.

6. *reseed_counter* = *reseed_counter* + 1.

7. Return **SUCCESS**, *returned_bits*, and the new values of V, C, and *reseed_counter* for the *new_working_state*.

Hashgen (*requested_number_of_bits*, V):

Input:

1. *requested_no_of_bits*: The number of bits to be returned.

2. *V*: The current value of V.

Output:

1. *returned_bits*: The generated bits to be returned to the generate function.

Hashgen Process:

1. $m = \left\lceil \dfrac{requested\ no\ of\ bits}{outlen} \right\rceil$.

2. *data* = V.

3. $W =$ the *Null* string.

4. For $i = 1$ to m

   4.1 $w_i =$ **Hash** (*data*).

   4.2 $W = W \| w_i$.

   4.3 $data = (data + 1) \bmod 2^{seedlen}$.

5. *returned_bits* = Leftmost (*requested_no_of_bits*) bits of W.

6. Return *returned_bits*.

10.1.2   HMAC_DRBG

HMAC_DRBG uses multiple occurrences of an **approved** keyed hash function, which is based on an **approved** hash function. This DRBG mechanism uses the **HMAC_DRBG_Update** function specified in Section 10.1.2.2 and the **HMAC** function within the **HMAC_DRBG_Update** function as the derivation function during instantiation and reseeding. The same hash function **shall** be used throughout an **HMAC_DRBG** instantiation. The hash function used **shall** meet or exceed the security requirements of the consuming application.

Figure 9 depicts the **HMAC_DRBG** in three stages. **HMAC_DRBG** is specified using an internal function (**HMAC_DRBG_Update**). This function is called during the **HMAC_DRBG** instantiate, generate and reseed algorithms to adjust the internal state when new entropy or additional input is provided, as well as to update the internal state after pseudorandom bits are generated. The operations in the top portion of the figure are only performed if the additional input is not null. Figure 10 depicts the **HMAC_DRBG_Update** function.

10.1.2.1   HMAC_DRBG Internal State

The internal state for **HMAC_DRBG** consists of:

1. The *working_state*:

    a. The value *V* of *outlen* bits, which is updated each time another *outlen* bits of output are produced (where *outlen* is specified in Table 2 of Section 10.1).

    b. The *outlen*-bit *Key*, which is updated at least once each time that the DRBG mechanism generates pseudorandom bits.

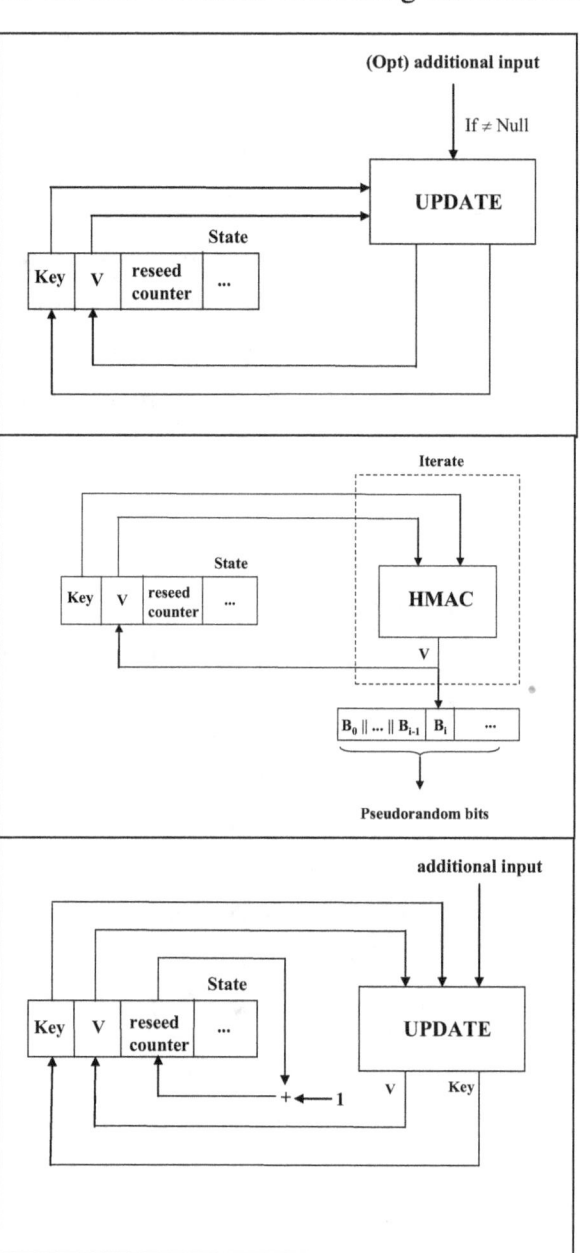

Figure 9: HMAC_DRBG Generate Function

c. A counter (*reseed_counter*) that indicates the number of requests for pseudorandom bits since instantiation or reseeding.

2. Administrative information:

    a. The *security_strength* of the DRBG instantiation.

    b. A *prediction_resistance_flag* that indicates whether or not a prediction resistance capability is required for the DRBG instantiation.

The value of *V* and *Key* are the critical values of the internal state upon which the security of this DRBG mechanism depends (i.e., *V* and *Key* are the "secret values" of the internal state).

10.1.2.2 The HMAC_DRBG Update Function (Update)

The **HMAC_DRBG_Update** function updates the internal state of **HMAC_DRBG** using the *provided_data*. Note that for this DRBG mechanism, the **HMAC_DRBG_Update** function also serves as a derivation function for the instantiate and reseed functions.

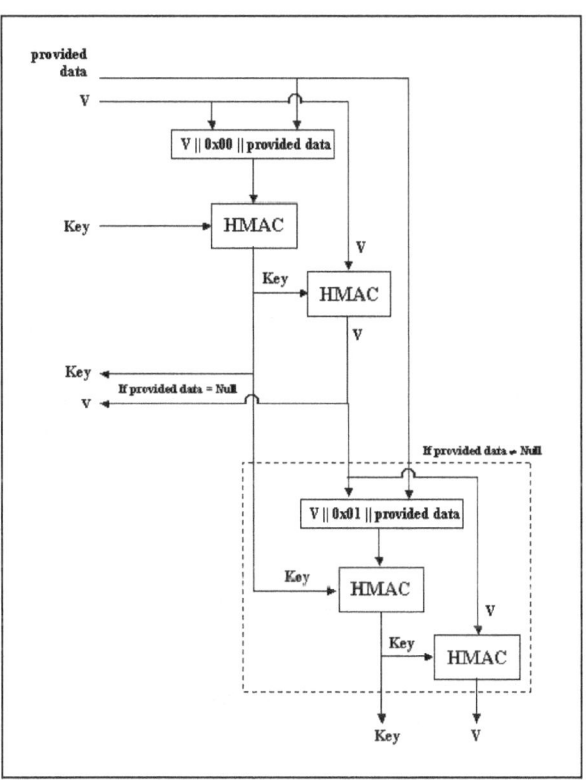

Figure 10: HMAC_DRBG_Update Function

Let **HMAC** be the keyed hash function specified in [FIPS 198] using the hash function selected for the DRBG mechanism from Table 2 in Section 10.1.

The following or an equivalent process **shall** be used as the **HMAC_DRBG_Update** function.

HMAC_DRBG_Update (*provided_data, K, V***):**

    1. *provided_data*: The data to be used.

    2. *K*: The current value of *Key*.

    3. *V*: The current value of *V*.

Output:

    1. *K*: The new value for *Key*.

    2. *V*: The new value for *V*.

HMAC_DRBG Update Process:

1. $K = \mathbf{HMAC}\ (K, V \parallel 0x00 \parallel provided\_data)$.

2. $V = \mathbf{HMAC}\ (K, V)$.

3. If ($provided\_data = Null$), then return K and V.

4. $K = \mathbf{HMAC}\ (K, V \parallel 0x01 \parallel provided\_data)$.

5. $V = \mathbf{HMAC}\ (K, V)$.

6. Return K and V.

10.1.2.3 Instantiation of HMAC_DRBG

Notes for the instantiate function specified in Section 9.1:

The instantiation of **HMAC_DRBG** requires a call to the **Instantiate_function** specified in Section 9.1. Process step 9 of that function calls the instantiate algorithm specified in this section. The values of *highest_supported_security_strength* and *min_length* are provided in Table 2 of Section 10.1. The contents of the internal state are provided in Section 10.1.2.1.

The instantiate algorithm:

Let **HMAC_DRBG_Update** be the function specified in Section 10.1.2.2. The output block length (*outlen*) is provided in Table 2 of Section 10.1.

The following process or its equivalent **shall** be used as the instantiate algorithm for this DRBG mechanism (see step 9 of the instantiate process in Section 9.1):

HMAC_DRBG_Instantiate_algorithm (*entropy_input, nonce, personalization_string, security_strength*)**:**

1. *entropy_input*: The string of bits obtained from the source of entropy input.

2. *nonce*: A string of bits as specified in Section 8.6.7.

3. *personalization_string*: The personalization string received from the consuming application. Note that the length of the *personalization_string* may be zero.

4. *security_strength*: The security strength for the instantiation. This parameter is optional for **HMAC_DRBG**, since it is not used.

Output:

1. *initial_working_state*: The inital values for *V*, *Key* and *reseed_counter* (see Section 10.1.2.1).

HMAC_DRBG Instantiate Process:

1. *seed_material = entropy_input $\parallel$ nonce $\parallel$ personalization_string*.

2. *Key* = 0x00 00...00.   Comment: *outlen* bits.

3. *V* = 0x01 01...01.   Comment: *outlen* bits.

          Comment: Update *Key* and *V*.

4. (*Key, V*) = **HMAC_DRBG_Update** (*seed_material, Key, V*).

5. *reseed_counter* = 1.

6. Return *V, Key* and *reseed_counter* as the *initial_working_state*.

10.1.2.4   Reseeding an HMAC_DRBG Instantiation

Notes for the reseed function specified in Section 9.2:

The reseeding of an **HMAC_DRBG** instantiation requires a call to the **Reseed_function** specified in Section 9.2. Process step 6 of that function calls the reseed algorithm specified in this section. The values for *min_length* are provided in Table 2 of Section 10.1.

The reseed algorithm:

Let **HMAC_DRBG_Update** be the function specified in Section 10.1.2.2. The following process or its equivalent **shall** be used as the reseed algorithm for this DRBG mechanism (see step 6 of the reseed process in Section 9.2):

HMAC_DRBG_Reseed_algorithm (*working_state, entropy_input, additional_input*):

1. *working_state*: The current values for *V, Key* and *reseed_counter* (see Section 10.1.2.1).

2. *entropy_input*: The string of bits obtained from the source of entropy input.

3. *additional_input*: The additional input string received from the consuming application. Note that the length of the *additional_input* string may be zero.

Output:

1. *new_working_state*: The new values for *V, Key* and *reseed_counter*.

HMAC_DRBG Reseed Process:

1. *seed_material* = *entropy_input* ‖ *additional_input*.

2. (*Key, V*) = **HMAC_DRBG_Update** (*seed_material, Key, V*).

3. *reseed_counter* = 1.

4. **Return** *V, Key* and *reseed_counter* as the *new_working_state*.

10.1.2.5   Generating Pseudorandom Bits Using HMAC_DRBG

Notes for the generate function specified in Section 9.3:

The generation of pseudorandom bits using an **HMAC_DRBG** instantiation requires a call to the **Generate_function** specified in Section 9.3. Process step 8 of that function calls the generate algorithm specified in this section. The values for *max_number_of_bits_per_request* and *outlen* are provided in Table 2 of Section 10.1.

The generate algorithm :

Let **HMAC** be the keyed hash function specified in [FIPS 198] using the hash function selected for the DRBG mechanism. The value for *reseed_interval* is defined in Table 2 of Section 10.1.

The following process or its equivalent **shall** be used as the generate algorithm for this DRBG mechanism (see step 8 of the generate process in Section 9.3):

HMAC_DRBG_Generate_algorithm (*working_state, requested_number_of_bits, additional_input*):

1. *working_state*: The current values for *V, Key* and *reseed_counter* (see Section 10.1.2.1).

2. *requested_number_of_bits*: The number of pseudorandom bits to be returned to the generate function.

3. *additional_input*: The additional input string received from the consuming application. Note that the length of the *additional_input* string may be zero.

Output:

1. *status*: The status returned from the function. The *status* will indicate **SUCCESS,** or indicate that a reseed is required before the requested pseudorandom bits can be generated.

2. *returned_bits*: The pseudorandom bits to be returned to the generate function.

3. *new_working_state*: The new values for *V, Key* and *reseed_counter.*

HMAC_DRBG Generate Process:

1. If *reseed_counter > reseed_interval*, then return an indication that a reseed is required.

2. If *additional_input ≠ Null*, then (*Key, V*) = **HMAC_DRBG_Update** (*additional_input, Key, V*).

3. *temp = Null.*

4. While (**len** (*temp*) < *requested_number_of_bits*) do:

    4.1   *V* = **HMAC** (*Key, V*).

    4.2   *temp = temp* ‖ *V.*

5. *returned_bits* = Leftmost *requested_number_of_bits* of *temp.*

6. (*Key, V*) = **HMAC_DRBG_Update** (*additional_input, Key, V*).

7. *reseed_counter = reseed_counter* + 1.

8. Return **SUCCESS**, *returned_bits*, and the new values of *Key, V* and *reseed_counter* as the *new_working_state*).

10.2 DRBG Mechanisms Based on Block Ciphers

A block cipher DRBG is based on a block cipher algorithm. The block cipher DRBG mechanism specified in this Recommendation has been designed to use any **approved** block cipher algorithm and may be used by consuming applications requiring various security strengths, providing that the appropriate block cipher algorithm and key length are used, and sufficient entropy is obtained for the seed.

The maximum security strength that can be supported by each DRBG based on a block cipher is the security strength of the block cipher and key size used; the security strengths for the block ciphers and key sizes are provided in [SP 800-57].

10.2.1   CTR_DRBG

CTR_DRBG uses an **approved** block cipher algorithm in the counter mode as specified in [SP 800-38A]. The same block cipher algorithm and key length **shall** be used for all block cipher operations of this DRBG. The block cipher algorithm and key length **shall** meet or exceed the security requirements of the consuming application.

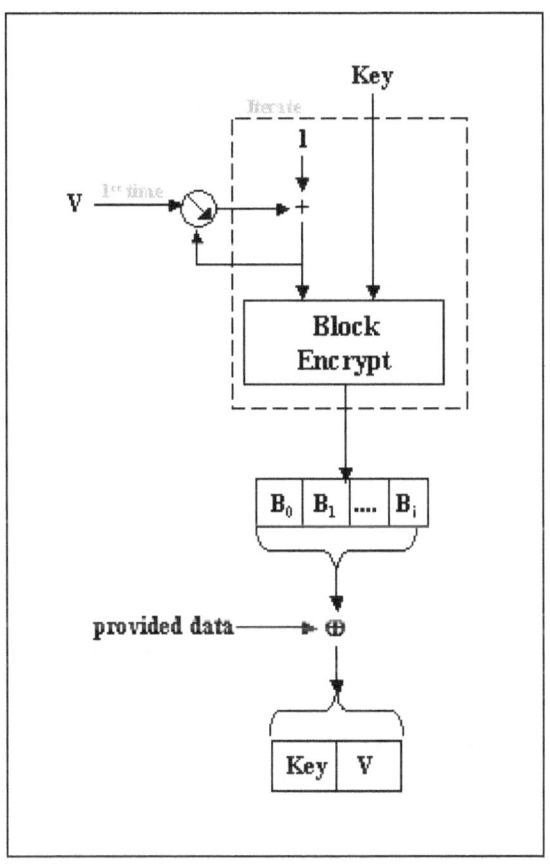

Figure 11: CTR_DRBG Update Function

CTR_DRBG is specified using an internal function (**CTR_DRBG_Update**). Figure 11 depicts the **CTR_DRBG_Update** function. This function is called by the instantiate, generate and reseed algorithms to adjust the internal state when new entropy or additional input is provided, as well as to update the internal state after pseudorandom bits are generated. Figure 12 depicts the **CTR_DRBG** in three stages. The operations in the top portion of the figure are only performed if the additional input is not null.

Table 3 specifies the values that **shall** be used for the function envelopes and **CTR_DRBG** mechanism

Table 3: Definitions for the CTR_DRBG

	3 Key TDEA	**AES-128**	**AES-192**	**AES-256**
Supported security strengths	See [SP 800-57]			

	3 Key TDEA	AES-128	AES-192	AES-256
highest_supported_security_strength	See [SP 800-57]			
Output block length (*outlen*)	64	128	128	128
Key length (*keylen*)	168	128	192	256
Required minimum entropy for instantiate and reseed	*security_strength*			
Seed length (*seedlen = outlen + keylen*)	232	256	320	384
If a derivation function is used:				
Minimum entropy input length (*min _length*)	*security_strength*			
Maximum entropy input length (*max _length*)	$\leq 2^{35}$ bits			
Maximum personalization string length (*max_personalization_string_length*)	$\leq 2^{35}$ bits			
Maximum additional_input length (*max_additional_input_length*)	$\leq 2^{35}$ bits			
If a derivation function is not used:				
Minimum entropy input length (*min _length = outlen + keylen*)	*seedlen*			
Maximum entropy input length (*max _length*) (*outlen + keylen*)	*seedlen*			
Maximum personalization string length (*max_personalization_string_length*)	*seedlen*			
Maximum additional_input length (*max_additional_input_length*)	*seedlen*			
max_number_of_bits_per_request	$\leq 2^{13}$		$\leq 2^{19}$	
Number of requests between reseeds (*reseed_interval*)	$\leq 2^{32}$		$\leq 2^{48}$	

The **CTR_ DRBG** may be implemented to use the block cipher derivation function specified in Section 10.4.2 during instantiation and reseeding. However, the DRBG mechanism is specified to allow an implementation tradeoff with respect to the use of this derivation function. The use of the derivation function is optional if either an **approved** RBG or an entropy source provide full entropy output when entropy input is requested by the DRBG mechanism.

Otherwise, the derivation function **shall** be used. Table 3 provides the lengths required for the *entropy_input*, *personalization_string* and *additional_input* for each case.

When using TDEA as the selected block cipher algorithm, the keys **shall** be handled as 64-bit blocks containing 56 bits of key and 8 bits of parity as specified for the TDEA engine specified in [SP 800-67].

10.2.1.1   CTR_DRBG Internal State

The internal state for the **CTR_DRBG** consists of:

1. The *working_state*:

   a. The value *V* of *outlen* bits, which is updated each time another *outlen* bits of output are produced.

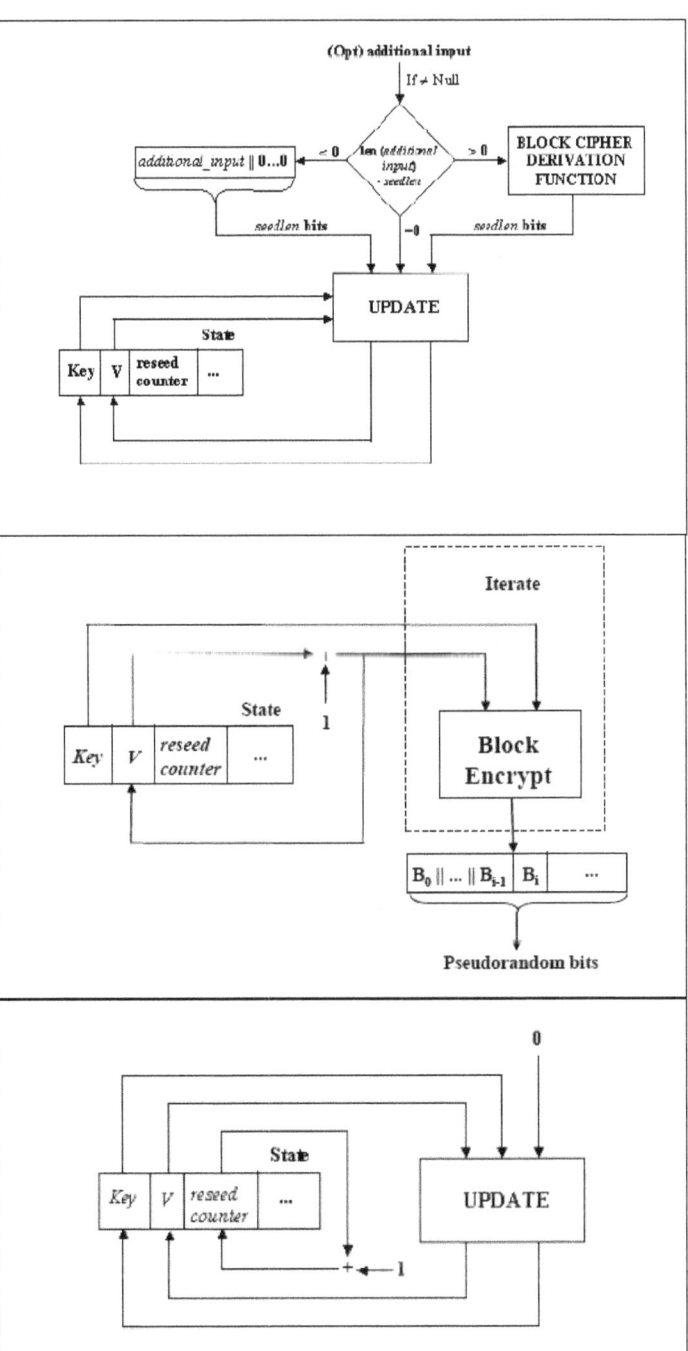

Figure 12: CTR-DRBG

     b. The *keylen*-bit *Key*, which is updated whenever a predetermined number of output blocks are generated.

     c. A counter (*reseed_counter*) that indicates the number of requests for pseudorandom bits since instantiation or reseeding.

  2. Administrative information:

     a. The *security_strength* of the DRBG instantiation.

     b. A *prediction_resistance_flag* that indicates whether or not a prediction resistance capability is required for the DRBG instantiation.

The values of *V* and *Key* are the critical values of the internal state upon which the security of this DRBG mechanism depends (i.e., *V* and *Key* are the "secret values" of the internal state).

10.2.1.2   The Update Function (CTR_DRBG_Update)

The **CTR_DRBG_Update** function updates the internal state of the **CTR_DRBG** using the *provided_data*. The values for *outlen*, *keylen* and *seedlen* are provided in Table 3 of Section 10.2.1. The block cipher operation in step 2.2 of the **CTR_DRBG_UPDATE** process uses the selected block cipher algorithm (also see Section 10.4). Note: the meaning of **Block_Encrypt** is discussed in Section 10.4.3.

The following or an equivalent process **shall** be used as the **CTR_DRBG_Update** function.

  CTR_DRBG_Update (*provided_data, Key, V*):

    1. *provided_data*: The data to be used. This must be exactly *seedlen* bits in length; this length is guaranteed by the construction of the *provided_data* in the instantiate, reseed and generate functions.

    2. *Key*: The current value of *Key*.

    3. *V*: The current value of *V*.

Output:

    1. *K*: The new value for *Key*.

    2. *V*: The new value for *V*.

CTR_DRBG_Update Process:

    1. *temp = Null*.

    2. While (**len** (*temp*) < *seedlen*) do

        2.1 $V = (V + 1) \bmod 2^{outlen}$.

        2.2 *output_block* = **Block_Encrypt** (*Key, V*).

        2.3 *temp = temp* ‖ *ouput_block*.

3. *temp* = Leftmost *seedlen* bits of *temp*.

4   *temp* = *temp* ⊕ *provided_data*.

5. *Key* = Leftmost *keylen* bits of *temp*.

6. *V* = Rightmost *outlen* bits of *temp*.

7. Return the new values of *Key* and V.

10.2.1.3   Instantiation of CTR_DRBG

Notes for the instantiate function specified in Section 9.1:

> The instantiation of **CTR_DRBG** requires a call to the **Instantiate_function**
> specified in Section 9.1. Process step 9 of that function calls the instantiate algorithm
> specified in this section. The values of *highest_supported_security_strength* and *min
> _length* are provided in Table 3 of Section 10.2.1. The contents of the internal state
> are provided in Section 10.2.1.1.

The instantiate algorithm:

> For this DRBG mechanism, there are two cases for processing. In each case, let
> **CTR_DRBG_Update** be the function specified in Section 10.2.1.2. The output block
> length (*outlen*), key length (*keylen*), seed length (*seedlen*) and *security_strengths* for
> the block cipher algorithms are provided in Table 3 of Section 10.2.1.

10.2.1.3.1 Instantiation When Full Entropy Input is Available, and a Derivation Function is Not Used

The following process or its equivalent **shall** be used as the instantiate algorithm for this
DRBG mechanism:

CTR_DRBG_Instantiate_algorithm (*entropy_input, personalization_string, security_strength*)**:**

1. *entropy_input*: The string of bits obtained from the source of entropy input.

2. *personalization_string*: The personalization string received from the
    consuming application. Note that the length of the *personalization_string* may
    be zero.

3. *security_strength*: The security strength for the instantiation. This parameter is
    optional for **CTR_DRBG**.

Output:

1. *initial_working_state*: The inital values for *V*, *Key*, and *reseed_counter* (see
    Section 10.2.1.1).

CTR_DRBG Instantiate Process:

1. *temp* = **len** (*personalization_string*).

> Comment: Ensure that the length of the *personalization_string* is exactly *seedlen* bits. The maximum length was checked in Section 9.1, processing step 3, using Table 3 to define the maximum length.

2. If (*temp* < *seedlen*), then *personalization_string* = *personalization_string* || $0^{seedlen - temp}$.

3. *seed_material* = *entropy_input* $\oplus$ *personalization_string*.

4. *Key* = 0^{keylen}.   Comment: *keylen* bits of zeros.

5. *V* = 0^{outlen}.   Comment: *outlen* bits of zeros.

6. (*Key*, *V*) = **CTR_DRBG_Update** (*seed_material*, *Key*, *V*).

7. *reseed_counter* = 1.

8. Return *V*, *Key*, and *reseed_counter* as the *initial_working_state*.

10.2.1.3.2   Instantiation When a Derivation Function is Used

Let **Block_Cipher_df** be the derivation function specified in Section 10.4.2 using the chosen block cipher algorithm and key size.

The following process or its equivalent **shall** be used as the instantiate algorithm for this DRBG mechanism:

CTR_DRBG_Instantiate_algorithm (*entropy_input, nonce, personalization_string, security_strength*):

1. *entropy_input*: The string of bits obtained from the source of entropy input.

2. *nonce*: A string of bits as specified in Section 8.6.7.

3. *personalization_string*: The personalization string received from the consuming application. Note that the length of the *personalization_string* may be zero.

4. *security_strength*: The security strength for the instantiation. This parameter is optional for **CTR_DRBG**, since it is not used.

Output:

1. *initial_working_state*: The inital values for *V*, *Key*, and *reseed_counter* (see Section 10.2.1.1).

CTR_DRBG Instantiate Process:

1. *seed_material* = *entropy_input* || *nonce* || *personalization_string*.

> Comment: Ensure that the length of the *seed_material* is exactly *seedlen* bits.

2. *seed_material* = **Block_Cipher_df** (*seed_material*, *seedlen*).

3. *Key* = 0^{keylen}.   Comment: *keylen* bits of zeros.

4. *V* = 0^{outlen}.   Comment: *outlen* bits of zeros.

5. (*Key*, *V*) = **CTR_DRBG_Update** (*seed_material*, *Key*, *V*).

6. *reseed_counter* = 1.

7. Return *V*, *Key*, and *reseed_counter* as the *initial_working_state*.

10.2.1.4   Reseeding a CTR_DRBG Instantiation

Notes for the reseed function specified in Section 9.2:

The reseeding of a **CTR_DRBG** instantiation requires a call to the **Reseed_function** specified in Section 9.2. Process step 6 of that function calls the reseed algorithm specified in this section. The values for *min _length* are provided in Table 3 of Section 10.2.1.

The reseed algorithm:

For this DRBG mechanism, there are two cases for processing. In each case, let **CTR_DRBG_Update** be the function specified in Section 10.2.1.2. The seed length (*seedlen*) is provided in Table 3 of Section 10.2.1.

10.2.1.4.1   Reseeding When Full Entropy Input is Available, and a Derivation Function is Not Used

The following process or its equivalent **shall** be used as the reseed algorithm for this DRBG mechanism (see step 6 of the reseed process in Section 9.2):

CTR_DRBG_Reseed_algorithm (*working_state*, *entropy_input*, *additional_input*):

1. *working_state*: The current values for *V*, *Key* and *reseed_counter* (see Section 10.2.1.1).

2. *entropy_input*: The string of bits obtained from the source of entropy input.

3. *additional_input*: The additional input string received from the consuming application. Note that the length of the *additional_input* string may be zero.

Output:

1. *new_working_state*: The new values for *V*, *Key*, and *reseed_counter*.

CTR_DRBG Reseed Process:

1. *temp* = **len** (*additional_input*).

Comment: Ensure that the length of the *additional_input* is exactly *seedlen* bits. The maximum length was checked in Section

9.2, processing step 2, using Table 3 to define the maximum length.

2. If (*temp* < *seedlen*), then *additional_input* = *additional_input* $\|$ $0^{seedlen - temp}$.

3. *seed_material* = *entropy_input* $\oplus$ *additional_input*.

4. (*Key*, *V*) = **CTR_DRBG_Update** (*seed_material*, *Key*, *V*).

5. *reseed_counter* = 1.

6. Return *V*, *Key* and *reseed_counter* as the *new_working_state*.

10.2.1.4.2   Reseeding When a Derivation Function is Used

Let **Block_Cipher_df** be the derivation function specified in Section 10.4.2 using the chosen block cipher algorithm and key size.

The following process or its equivalent **shall** be used as the reseed algorithm for this DRBG mechanism (see reseed process step 6 of Section 9.2):

CTR_DRBG_Reseed_algorithm (*working_state*, *entropy_input*, *additional_input*)

1. *working_state*: The current values for *V*, *Key* and *reseed_counter* (see Section 10.2.1.1).

2. *entropy_input*: The string of bits obtained from the source of entropy input.

3. *additional_input*: The additional input string received from the consuming application. Note that the length of the *additional_input* string may be zero.

Output:

1. *new_working_state*: The new values for *V*, *Key*, and *reseed_counter*.

CTR_DRBG Reseed Process:

1. *seed_material* = *entropy_input* $\|$ *additional_input*.

> Comment: Ensure that the length of the *seed_material* is exactly *seedlen* bits.

2. *seed_material* = **Block_Cipher_df** (*seed_material*, *seedlen*).

3. (*Key*, *V*) = **CTR_DRBG_Update** (*seed_material*, *Key*, *V*).

4. *reseed_counter* = 1.

5. Return *V*, *Key*, and *reseed_counter* as the *new_working_state*.

10.2.1.5   Generating Pseudorandom Bits Using CTR_DRBG

Notes for the generate function specified in Section 9.3:

The generation of pseudorandom bits using a **CTR_DRBG** instantiation requires a call to the **Generate_function** specified in Section 9.3. Process step 8 of that function calls the generate algorithm specified in this section. The values for *max_number_of_bits_per_request* and *max_additional_input_length*, and *outlen* are provided in Table 3 of Section 10.2.1. If the derivation function is not used, then the maximum allowed length of *additional_input* = *seedlen*.

For this DRBG mechanism, there are two cases for the processing. For each case, let **CTR_DRBG_Update** be the function specified in Section 10.2.1.2, and let **Block_Encrypt** be the function specified in Section 10.4.3. The seed length (*seedlen*) and the value of *reseed_interval* are provided in Table 3 of Section 10.2.1.

10.2.1.5.1   Generating Pseudorandom Bits When a Derivation Function is <ins>Not</ins> Used for the DRBG Implementation

The following process or its equivalent **shall** be used as the generate algorithm for this DRBG mechanism (see step 8 of the generate process in Section 9.3.3):

CTR_DRBG_Generate_algorithm (*working_state, requested_number_of_bits, additional_input*):

1. *working_state*: The current values for *V, Key*, and *reseed_counter* (see Section 10.2.1.1).

2. *requested_number_of_bits*: The number of pseudorandom bits to be returned to the generate function.

3. *additional_input*: The additional input string received from the consuming application. Note that the length of the *additional_input* string may be zero.

Output:

1. *status*: The status returned from the function. The *status* will indicate **SUCCESS,** or indicate that a reseed is required before the requested pseudorandom bits can be generated.

2. *returned_bits*: The pseudorandom bits returned to the generate function.

3. *working_state*: The new values for *V, Key*, and *reseed_counter*.

CTR_DRBG Generate Process:

1. If *reseed_counter* > *reseed_interval*, then return an indication that a reseed is required.

2. If (*additional_input* ≠ *Null*), then

> Comment: Ensure that the length of the *additional_input* is exactly *seedlen* bits. The maximum length was checked in Section 9.3.3, processing step 4, using Table 3 to

define the maximum length. If the length of the *additional input* is < *seedlen*, pad with zero bits.

  2.1   *temp* = **len** (*additional_input*).

  2.2   If (*temp* < *seedlen*), then
additional_input = *additional_input* $\|$ $0^{seedlen \, - \, temp}$.

  2.3   (*Key*, *V*) = **CTR_DRBG_Update** (*additional_input*, *Key*, *V*).

Else *additional_input* = $0^{seedlen}$.

3. *temp* = *Null*.

4. While (**len** (*temp*) < *requested_number_of_bits*) do:

  4.1 $V = (V + 1) \bmod 2^{outlen}$.

  4.2   *output_block* = **Block_Encrypt** (*Key*, V).

  4.3   *temp* = *temp* $\|$ *output_block*.

5. *returned_bits* = Leftmost *requested_number_of_bits* of *temp*.

> Comment: Update for backtracking resistance.

6. (*Key*, *V*) = **CTR_DRBG_Update** (*additional_input*, *Key*, *V*).

7. *reseed_counter* = *reseed_counter* + 1.

8. Return **SUCCESS** and *returned_bits*; also return *Key*, *V*, and *reseed_counter* as the *new_working_state*.

10.2.1.5.2   Generating Pseudorandom Bits When a Derivation Function <u>is</u> Used for the DRBG Implementation

The **Block_Cipher_df** is specified in Section 10.4.2 and **shall** be implemented using the chosen block cipher algorithm and key size.

The following process or its equivalent **shall** be used as the generate algorithm for this DRBG mechanism (see step 8 of the generate process in Section 9.3.3):

CTR_DRBG_Generate_algorithm (*working_state, requested_number_of_bits, additional_input*):

1. *working_state*: The current values for *V*, *Key*, and *reseed_counter* (see Section 10.2.1.1).

2. *requested_number_of_bits*: The number of pseudorandom bits to be returned to the generate function.

3. *additional_input*: The additional input string received from the consuming application. Note that the length of the *additional_input* string may be zero.

Output:

1. *status*: The status returned from the function. The *status* will indicate **SUCCESS,** or indicate that a reseed is required before the requested pseudorandom bits can be generated.

2. *returned_bits*: The pseudorandom bits returned to the generate function.

3. *working_state*: The new values for *V*, *Key*, and *reseed_counter*.

CTR_DRBG Generate Process:

1. If *reseed_counter > reseed_interval*, then return an indication that a reseed is required.

2. If (*additional_input ≠ Null*), then

    2.1   *additional_input* = **Block_Cipher_df** (*additional_input*, *seedlen*).

    2.2   (*Key*, *V*) = **CTR_DRBG_Update** (*additional_input*, *Key*, *V*).

    Else *additional_input* = $0^{seedlen}$.

3. *temp = Null*.

4. While (**len** (*temp*) < *requested_number_of_bits*) do:

    4.1 $V = (V + 1) \bmod 2^{outlen}$.

    4.2   *output_block* = **Block_Encrypt** (*Key*, V).

    4.3   *temp = temp* ‖ *output_block*.

5. *returned_bits* = Leftmost *requested_number_of_bits* of *temp*.

    > Comment: Update for backtracking resistance.

6. (*Key*, *V*) = **CTR_DRBG_Update** (*additional_input*, *Key*, *V*).

7. *reseed_counter* = *reseed_counter* + 1.

8. Return **SUCCESS** and *returned_bits*; also return *Key*, *V*, and *reseed_counter* as the *new_working_state*.

10.3   DRBG Mechanisms Based on Number Theoretic Problems

A DRBG can be designed to take advantage of number theoretic problems (e.g., the discrete logarithm problem). If done correctly, such a generator's properties of randomness and/or unpredictability will be assured by the difficulty of finding a solution to that problem. This section specifies a DRBG mechanism that is based on the elliptic curve discrete logarithm problem.

10.3.1   Dual Elliptic Curve Deterministic RBG (Dual_EC_DRBG)

Dual_EC_DRBG is based on the following hard problem, sometimes known as the "elliptic curve discrete logarithm problem" (ECDLP): given points P and Q on an elliptic curve of order n, find a such that $Q = aP$.

Dual_EC_DRBG uses an initial seed that is 2 * *security_strength* bits in length to initiate the generation of *outlen*-bit pseudorandom strings by performing scalar multiplications on two points in an elliptic curve group, where the curve is defined over a field approximately 2^m in size. For all the NIST curves given in this Recommendation, m is at least twice the *security_strength*, and never less than 256. Throughout this DRBG mechanism specification, m will be referred to as *seedlen*; the term "*seedlen*" is appropriate because the internal state of **Dual_EC_DRBG** is used as a "seed" for the random block it produces. Figure 13 depicts the **Dual_EC_DRBG**.

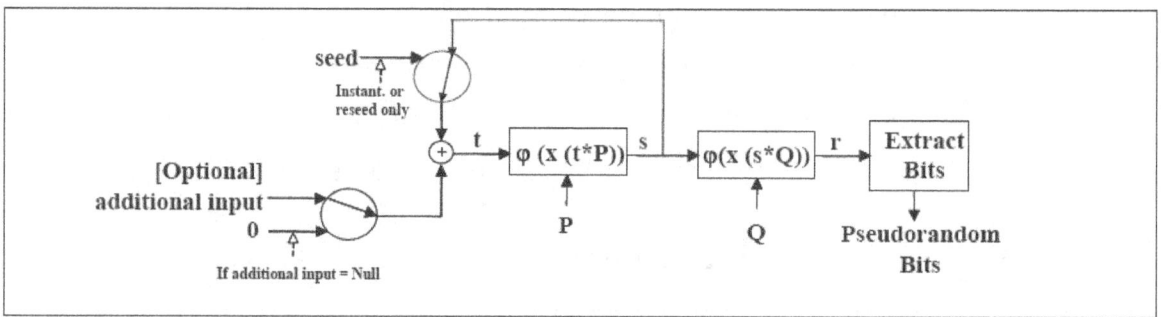

Figure 13: Dual_EC_DRBG

The instantiation of this DRBG mechanism requires the selection of an appropriate elliptic curve and curve points specified in Appendix A.1 for the desired security strength. The *seed* used to determine the initial value (s) of the DRBG mechanism **shall** have at least *security_strength* bits of entropy. Further requirements for the *seed* are provided in Section 8.6. This DRBG mechanism uses the derivation function specified in Section 10.4.1 during instantiation and reseeding.

The maximum security strength that can be supported by the **Dual_EC_DRBG** is the security strength of the curve used; the security strengths for the curves are provided in [SP 800-57].

Backtracking resistance is inherent in the algorithm, even if the internal state is compromised. As shown in Figure 14, **Dual_EC_DRBG** generates a *seedlen*-bit number for each step $i = 1, 2, 3, \ldots,$ as follows:

$$s_i = \varphi(\, x(s_{i-1} * P)\,)$$

$$r_i = \varphi(\, x(s_i * Q)\,).$$

Each arrow in the figure represents an Elliptic Curve scalar multiplication operation, followed by the extraction of the x coordinate for the resulting point and for the random output r_i, followed by truncation to produce the output. Following a line in the direction of the arrow is the normal operation; inverting the direction implies the ability to solve the ECDLP for that specific curve. An adversary's ability to invert an arrow in the figure implies that the adversary has solved the ECDLP for that specific elliptic curve. Backtracking resistance is built into the design, as knowledge of s_1 does not allow an adversary to determine s_0 (and so forth) unless the adversary is able to solve the ECDLP for that specific curve. In addition, knowledge of r_1 does not allow an adversary to determine s_1 (and so forth) unless the adversary is able to solve the ECDLP for that specific curve.

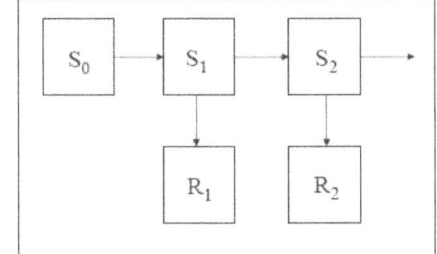

Figure 14: Dual_EC_DRBG Backtracking Resistance

Table 4 specifies the values that **shall** be used for the envelope and algorithm for each curve. Complete specifications for each curve are provided in Appendix A.1.

Table 4: Definitions for the Dual_EC_DRBG

	P-256	P-384	P-521
Supported security strengths	See [SP 800-57]		
Size of the base field (in bits), referenced throughout as *seedlen*	256	384	521
highest_supported_ security_strength	See [SP 800-57]		
Output block length (*max_outlen* = largest multiple of 8 less than (size of the base field) - (13 + log₂ (the cofactor))	240	368	504
Required minimum entropy for instantiate and reseed	*security_strength*		
Minimum entropy input length (*min_length*)	*security_strength*		
Maximum entropy input length (*max_length*)	$\leq 2^{13}$ bits		
Maximum personalization string length (*max_personalization_string_length*)	$\leq 2^{13}$ bits		

	P-256	P-384	P-521
Maximum additional input length (***max_additional_input_length***)	$\le 2^{13}$ bits		
Length of the initial seed	$2 \times security\_strength$		
Appropriate hash functions	SHA-1, SHA-224, SHA-512/224, SHA-256, SHA-512/256, SHA-384, SHA-512	SHA-224, SHA-512/224, SHA-256, SHA-512/256, SHA-384, SHA-512	SHA-256, SHA-512/256, SHA-384, SHA-512
max_number_of_bits_per_request	$max\_outlen \times reseed\_interval$		
Number of blocks between reseeding (***reseed_interval***)	$\le 2^{32}$ blocks		

10.3.1.1   Dual_EC_DRBG Internal State

The internal state for **Dual_EC_DRBG** consists of:

1. The *working_state*:

   a. A value (*s*) that determines the current position on the curve.

   b. The elliptic curve domain parameters (*seedlen, p, a, b, n*), where *seedlen* is the length of the seed, *p* is the prime that defines the base field F_p, *a* and *b* are two field elements that define the equation of the curve, and *n* is the order of the point *G*. If only one curve will be used by an implementation, these parameters need not be present in the *working_state*.

   c. Two points *P* and *Q* on the curve (see Appendix A) will be used. If only one curve will be used by an implementation, these points need not be present in the *working_state*.

   d. A counter (*reseed_counter*) that indicates the number of blocks of random data produced by the **Dual_EC_DRBG** since the initial seeding or the previous reseeding.

2. Administrative information:

   a. The *security_strength* provided by the DRBG instantiation,

   b. A *prediction_resistance_flag* that indicates whether prediction resistance is required by the DRBG instantiation.

The value of *s* is the critical value of the internal state upon which the security of this DRBG mechanism depends (i.e., *s* is the "secret value" of the internal state).

10.3.1.2   Instantiation of Dual_EC_DRBG

Notes for the instantiate function specified in Section 9.1:

The instantiation of **Dual_EC_DRBG** requires a call to the instantiate function specified in Section 9.1. Process step 9 of that function calls the instantiate algorithm in this section.

When selecting the curve in step 4 below, it is recommended that the default values be used for P and Q as given in Appendix A.1. However, an implementation may use different pairs of points, provided that they are *verifiably random*, as evidenced by the use of the procedure specified in Appendix A.2.1 and the self-test procedure described in Appendix A.2.2.

The values for *highest_supported_security_strength* and *min_length* are determined by the selected curve (see Table 4 in Section 10.3.1).

The instantiate algorithm :

Let **Hash_df** be the hash derivation function specified in Section 10.4.1 using an appropriate hash function from Table 4 in Section 10.3.1. Let *seedlen* be the appropriate value from Table 4.

The following process or its equivalent **shall** be used as the instantiate algorithm for this DRBG mechanism (see step 9 of the instantiate process in Section 9.1):

Dual_EC_DRBG_Instantiate_algorithm (*entropy_input, nonce, personalization_string, security_strength*):

1. *entropy_input*: The string of bits obtained from the source of entropy input.

2. *nonce*: A string of bits as specified in Section 8.6.7.

3. *personalization_string*: The personalization string received from the consuming application. Note that the length of the *personalization string* may be zero.

4. *security_strength*: The security strength for the instantiation. This parameter is required for **Dual_EC_DRBG**.

Output:

1. *s, seedlen, p, a, b, n, P, Q*, and a *reseed_counter* for the *initial_working_state*.

Dual_EC_DRBG Instantiate Process:

1. *seed_material = entropy_input* || *nonce* || *personalization_string*.

    > Comment: Use a hash function to ensure that the entropy is distributed throughout the bits, and *s* is *m* (i.e., *seedlen*) bits in length.

2. *s* = **Hash_df** (*seed_material, seedlen*).

3. *reseed_counter* = 0.

4. Using the *security strength* and Table 4 in Section 10.3.1, select the smallest available curve that has a security strength $\geq$ *security_strength*. The values for *seedlen, p, a, b, n, P, Q* are determined by the curve.

5. Return *s, seedlen, p, a, b, n, P, Q,* and a *reseed_counter* for the *initial_working_state*.

10.3.1.3   Reseeding of a Dual_EC_DRBG Instantiation

Notes for the reseed function specified in Section 9.2:

The reseed of **Dual_EC_DRBG** requires a call to the reseed function specified in Section 9.2. Process step 6 of that function calls the reseed algorithm in this section. The values for *min_length* are provided in Table 4 of Section 10.3. 1.

The reseed algorithm :

Let **Hash_df** be the hash derivation function specified in Section 10.4.1 using an appropriate hash function from Table 4 in Section 10.3. 1.

The following process or its equivalent **shall** be used to reseed the **Dual_EC_DRBG** process after it has been instantiated (see step 6 of the reseed process in Section 9.2):

Dual_EC_DRBG_Reseed_algorithm (*working_state, entropy_input, additional_input*)**:**

1. *working_state*: The current values for *s, seedlen, p, a, b, n, P, Q,* and a *reseed_counter* (see Section 10.3.1.1).

2. *entropy_input*: The string of bits obtained from the source of entropy input.

3. *additional_input*: The additional input string received from the consuming application. Note that the length of the *additional_input* string may be zero.

Output:

1. *s, seedlen, p, a, b, n, P, Q,* and a *reseed_counter* for the *new_working_state*.

Dual_EC_DRBG Reseed Process:

> Comment: **pad8** returns a copy of *s* padded on the right with binary 0's, if necessary, to a multiple of 8.

1. *seed_material* = **pad8** (*s*) ∥ *entropy_input* ∥ *additional_input*.

2. *s* = **Hash_df** (*seed_material, seedlen*).

3. *reseed_counter* = 0.

4. Return *s, seedlen, p, a, b, n, P, Q,* and a *reseed_counter* for the *new_working_state*.

10.3.1.4   Generating Pseudorandom Bits Using Dual_EC_DRBG

Notes for the generate function specified in Section 9.3:

The generation of pseudorandom bits using a **Dual_EC_DRBG** instantiation requires a call to the generate function. Process step 8 of that function calls the generate algorithm specified in this section. The values for *max_number_of_bits_per_request* and *max_outlen* are provided in Table 4 of Section 10.3.1. *outlen* is the number of pseudorandom bits taken from each *x*-coordinate as the **Dual_EC_DRBG** steps. For performance reasons, the value of *outlen* **should** be set to the maximum value as provided in Table 4. If the value of *outlen* provided in Table 4 is not used, then *outlen* **shall** be set to any multiple of eight bits less than or equal to *max_outlen*. The bits that become the **Dual_EC_DRBG** output are always the rightmost bits, i.e., the least significant bits of the *x*-coordinates. Appendix E.2 contains additional information regarding the statistical and distributional implications related to the truncation of the *x*-coordinates.

The generate algorithm:

Let **Hash_df** be the hash derivation function specified in Section 10.4.1 using an appropriate hash function from Table 4 in Section 10.3.1. The value of *reseed_interval* is also provided in Table 4.

The following are used by the generate algorithm:

a. **pad8** (*bitstring*) returns a copy of the *bitstring* padded on the right with binary 0's, if necessary, to a multiple of 8.

b. **Truncate** (*bitstring, in_len, out_len*) inputs a *bitstring* of *in_len* bits, returning a string consisting of the leftmost *out_len* bits of *bitstring*. If *in_len* < *out_len*, the *bitstring* is padded on the right with (*out_len* - *in_len*) zeroes, and the result is returned.

c. *x*(*A*) is the *x*-coordinate of the point *A* on the curve, given in affine coordinates. An implementation may choose to represent points internally using other coordinate systems; for instance, when efficiency is a primary concern. In this case, a point **shall** be translated back to affine coordinates before *x*() is applied.

d. φ (*x*) maps field elements to non-negative integers, taking the bit vector representation of a field element and interpreting it as the binary expansion of an integer.

The precise definition of $\varphi(x)$ used in steps 6 and 7 of the generate process below depends on the field representation of the curve points. In keeping with the convention of [FIPS 186], the following elements will be associated with each other (note that, in this case, *m* denotes the size of the base field):

B: $|c_{m-1}|c_{m-2}|\ ...\ |c_1|c_0|$, a bitstring, with c_{m-1} being leftmost

Z: $c_{m-1}2^{m-1} + ... + c_2 2^2 + c_1 2^1 + c_0 \in Z$;

Fa: $c_{m-1}2^{m-1} + ... + c_2 2^2 + c_1 2^1 + c_0 \bmod p \in F_p$;

Thus, any field element *x* of the form F_a will be converted to the integer *Z* or bitstring *B*, and vice versa, as appropriate.

e. * is the symbol representing scalar multiplication of a point on the curve.

The following process or its equivalent **shall** be used to generate pseudorandom bits (see step 8 of the generate process in Section 9.3):

Dual_EC_DRBG_Generate_algorithm (*working_state, requested_number_of_bits, additional_input*):

1. *working_state*: The current values for *s, seedlen, p, a, b, n, P, Q,* and a *reseed_counter* (see Section 10.3.1.1).

2. *requested_number_of_bits*: The number of pseudorandom bits to be returned to the generate function.

3. *additional_input*: The additional input string received from the consuming application. If the input of *additional_input* is not supported by an implementation, then step 2 of the generate process becomes:

    additional_input = 0.

    Alternatively, generate steps 2 and 9 are omitted, the *additional_input* term is omitted from step 5, and the "go to step 5" in step 12 is to the step that now sets *t = s*.

Output:

1. *status*: The status returned from the function. The *status* will indicate **SUCCESS,** or an indication that a reseed is required before the requested pseudorandom bits can be generated.

2. *returned_bits*: The pseudorandom bits to be returned to the generate function.

3. *s, seedlen, p, a, b, n, P, Q,* and a *reseed_counter* for the *new_working_state*.

Dual_EC_DRBG Generate Process:

> Comment: Check whether a reseed is required.

1. If $\left(reseed\_counter + \left\lceil \dfrac{requested\_no\_of\_bits}{outlen} \right\rceil \right) > reseed\_interval$, then return an indication that a reseed is required.

> Comment: If *additional_input* is *Null*, set to *seedlen* zeroes; otherwise, **Hash_df** to *seedlen* bits.

2. If (*additional_input_string = Null*), then *additional_input* = 0

    Else   *additional_input* = **Hash_df** (**pad8** (*additional_input_string*), *seedlen*).

> Comment: Produce *requested_no_of_bits*, *outlen* bits at a time:

3. *temp* = the *Null* string.

4 *i* = 0.

5. $t = s \oplus additional\_input$.

    Comment: t is to be interpreted as a *seedlen*-bit unsigned integer. To be precise, t should be reduced mod n; the operation * will effect this.

6. $s = \varphi(x(t * P))$.

    Comment: s is a *seedlen*-bit number. Note that the conversion of $\varphi(x)$ is discussed in item d above; this also applies to step 7.

7. $r = \varphi(x(s * Q))$.

    Comment: r is a *seedlen*-bit number.

8. $temp = temp \,\|\,$ (**rightmost** *outlen* bits of r).

9. *additional_input*=0

    Comment: *seedlen* zeroes; *additional_input_string* is added only on the first iteration.

10. *reseed_counter* = *reseed_counter* + 1.

11. $i = i + 1$.

12. If (**len** (*temp*) < *requested_number_of_bits*), then go to step 5.

13. *returned_bits* = **Truncate** (*temp*, $i \times$ *outlen*, *requested_number_of_bits*).

14. $s = \varphi(x(s * P))$.

15. Return **SUCCESS**, *returned_bits*, and s, *seedlen*, p, a, b, n, P, Q, and a *reseed_counter* for the *new_working_state*.

10.4   Auxiliary Functions

Derivation functions are internal functions that are used during DRBG instantiation and reseeding to either derive internal state values or to distribute entropy throughout a bitstring. Two methods are provided. One method is based on hash functions (see Section 10.4.1), and the other method is based on block cipher algorithms (see 10.4.2). The block cipher derivation function specified in Section 10.4.2 uses a **Block_Cipher_Hash** function that is specified in Section 10.4.3.

The presence of these derivation functions in this Recommendation does not implicitly approve these functions for any other application.

10.4.1   Derivation Function Using a Hash Function (Hash_df)

This derivation function is used by the **Hash_DRBG** and **Dual_EC_DRBG** specified Section 10.1.1 and 10.3.1, respectively. The hash-based derivation function hashes an input string and returns the requested number of bits. Let **Hash** be the hash function used by the DRBG mechanism, and let *outlen* be its output length.

The following or an equivalent process **shall** be used to derive the requested number of bits.

Hash_df (*input_string, no_of_bits_to_return*)**:**

1. *input_string*: The string to be hashed.

2. *no_of_bits_to_return*: The number of bits to be returned by **Hash_df.** The maximum length (*max_number_of_bits*) is implementation dependent, but **shall** be less than or equal to (255 × *outlen*). *no_of_bits_to_return* is represented as a 32-bit integer.

Output:

1. *status*: The status returned from **Hash_df**. The status will indicate **SUCCESS** or **ERROR_FLAG**.

2. *requested_bits* : The result of performing the **Hash_df**.

Hash_df Process:

1. *temp* = the Null string.

2. $len = \left\lceil \dfrac{no\_of\_bits\_to\_return}{outlen} \right\rceil.$

3. *counter* = an 8-bit binary value representing the integer "1".

4. For *i* = 1 to *len* do

> Comment : In step 4.1, *no_of_bits_to_return* is used as a 32-bit string.

    4.1   *temp* = *temp* || **Hash** (*counter* || *no_of_bits_to_return* || *input_string*).

    4.2   *counter* = *counter* + 1.

5. *requested_bits* = Leftmost (*no_of_bits_to_return*) of *temp*.

6. Return **SUCCESS** and *requested_bits*.

10.4.2   Derivation Function Using a Block Cipher Algorithm (Block_Cipher_df)

This derivation function is used by the **CTR_DRBG** that is specified in Section 10.2. Let **BCC** be the function specified in Section 10.4.3. Let *outlen* be its output block length, which is a multiple of 8 bits for the **approved** block cipher algorithms, and let *keylen* be the key length.

The following or an equivalent process **shall** be used to derive the requested number of bits.

Block_Cipher_df (*input_string, no_of_bits_to_return*)**:**

1. *input_string*: The string to be operated on. This string **shall** be a multiple of 8 bits.

2. *no_of_bits_to_return*: The number of bits to be returned by **Block_Cipher_df**. The maximum length (*max_number_of_bits*) is 512 bits for the currently **approved** block cipher algorithms.

Output:

1. *status*: The status returned from **Block_Cipher_df**. The status will indicate **SUCCESS** or **ERROR FLAG**.

2. *requested_bits* : The result of performing the **Block_Cipher_df**.

Block_Cipher_df Process:

1. If (*number_of_bits_to_return* > *max_number_of_bits*), then return an **ERROR_FLAG**.

2. L = **len** (*input_string*)/8.
   Comment: L is the bitstring representation of the integer resulting from **len** (*input_string*)/8. L **shall** be represented as a 32-bit integer.

3. N = *number_of_bits_to_return*/8.
   Comment : N is the bitstring represention of the integer resulting from *number_of_bits_to_return*/8. N **shall** be represented as a 32-bit integer.

   Comment: Prepend the string length and the requested length of the output to the *input_string*.

4. $S = L \parallel N \parallel input\_string \parallel 0x80$.

   Comment : Pad S with zeros, if necessary.

5. While (**len** (S) mod *outlen*) $\neq$ 0, $S = S \parallel 0x00$.

   Comment : Compute the starting value.

6. *temp* = the *Null* string.

7. $i = 0$.
   Comment : i **shall** be represented as a 32-bit integer, i.e., **len** (i) = 32.

8. K = Leftmost *keylen* bits of 0x00010203...1D1E1F.

9. While **len** (*temp*) < *keylen* + *outlen*, do

   9.1 $IV = i \parallel 0^{outlen - \text{len} (i)}$.
   Comment: The 32-bit integer represenation of i is padded with zeros to *outlen* bits.

   9.2   *temp* = *temp* $\parallel$ **BCC** (K, ($IV \parallel S$)).

   9.3 $i = i + 1$.

   Comment: Compute the requested number of bits.

10. K = Leftmost *keylen* bits of *temp*.

11. X = Next *outlen* bits of *temp*.

12. *temp* = the *Null* string.

13. While **len** (*temp*) < *number_of_bits_to_return*, do

   13.1 *X* = **Block_Encrypt** (*K*, *X*).

   13.2 *temp* = *temp* || *X*.

14. *requested_bits* = Leftmost *number_of_bits_to_return* of *temp*.

15. Return **SUCCESS** and *requested_bits*.

10.4.3   BCC Function

Block_Encrypt is used for convenience in the specification of the **BCC** function. This function is not specifically defined in this Recommendation, but has the following meaning:

> **Block_Encrypt:** A basic encryption operation that uses the selected block cipher algorithm. The function call is:

$$output\_block = \textbf{Block\_Encrypt} (Key, input\_block)$$

> For TDEA, the basic encryption operation is called the forward cipher operation (see [SP 800-67]); for AES, the basic encryption operation is called the cipher operation (see [FIPS 197]). The basic encryption operation is equivalent to an encryption operation on a single block of data using the ECB mode.

For the **BCC** function, let *outlen* be the length of the output block of the block cipher algorithm to be used.

The following or an equivalent process **shall** be used to derive the requested number of bits.

BCC (*Key, data*):

1. *Key*: The key to be used for the block cipher operation.

2. *data*: The data to be operated upon. Note that the length of *data* must be a multiple of *outlen*. This is guaranteed by **Block_Cipher_df** process steps 4 and 8.1 in Section 10.4.2.

Output:

1. *output_block*: The result to be returned from the **BCC** operation.

BCC Process:

1. *chaining_value* = 0^{outlen}.   Comment: Set the first chaining value to *outlen* zeros.

2. *n* = **len** (*data*)/*outlen*.

3. Starting with the leftmost bits of data, split the *data* into *n* blocks of *outlen* bits each, forming $block_1$ to $block_n$.

4. For *i* = 1 to *n* do

   4.1 *input_block* = *chaining_value* ⊕ $block_i$.

4.2   *chaining_value* = **Block_Encrypt** (*Key*, *input_block*).

5. *output_block* = chaining_value.

6. Return *output_block*.

11 Assurance

A user of a DRBG for cryptographic purposes requires assurance that the generator actually produces (pseudo) random and unpredictable bits. The user needs assurance that the design of the generator, its implementation and its use to support cryptographic services are adequate to protect the user's information. In addition, the user requires assurance that the generator continues to operate correctly.

The design of each DRBG mechanism in this Recommendation has received an evaluation of its security properties prior to its selection for inclusion in this Recommendation.

An implementation **shall** be validated for conformance to this Recommendation by a NVLAP-accredited laboratory (see Section 11.2). Such validations provide a higher level of assurance that the DRBG mechanism is correctly implemented. Validation testing for DRBG mechanisms consists of testing whether or not the DRBG mechanism produces the expected result, given a specific set of input parameters (e.g., entropy input).

Health tests on the DRBG mechanism **shall** be implemented within a DRBG mechanism boundary or sub-boundary in order to determine that the process continues to operate as designed and implemented. See Section 11.3 for further information.

A cryptographic module containing a DRBG mechanism **shall** be validated (see [FIPS 140]). The consuming application or cryptographic service that uses a DRBG mechanism **should** also be validated and periodically tested for continued correct operation. However, this level of testing is outside the scope of this Recommendation.

Note that any entropy input used for testing (either for validation testing or health testing) may be publicly known. Therefore, entropy input used for testing **shall not** knowingly be used for normal operational use.

11.1   Minimal Documentation Requirements

A set of documentation **shall** be developed that will provide assurance to users and validators that the DRBG mechanisms in this Recommendation have been implemented properly. Much of this documentation may be placed in a user's manual. This documentation **shall** consist of the following as a minimum:

- Document the method for obtaining entropy input.

- Document how the implementation has been designed to permit implementation validation and health testing.

- Document the type of DRBG mechanism (e.g., **CTR_DRBG**, **Dual_EC_DRBG**), and the cryptographic primitives used (e.g., AES-128, SHA-256).

- Document the security strengths supported by the implementation.

- Document features supported by the implementation (e.g., prediction resistance, the available elliptic curves, etc.).

- If DRBG mechanism functions are distributed, specify the mechanisms that are used to protect the confidentiality and integrity of the internal state or parts of the internal state that are transferred between the distributed DRBG mechanism sub-boundaries (i.e., provide documentation about the secure channel).

- In the case of the **CTR_DRBG**, indicate whether a derivation function is provided. If a derivation function is not used, document that the implementation can only be used when full entropy input is available.

- Document any support functions other than health testing.

- Document the periodic intervals at which health testing is performed for the generate function and provide a justification for the selected intervals (see Section 11.3.3).

- Document how the integrity of the health tests will be determined subsequent to implementation validation testing.

11.2   Implementation Validation Testing

A DRBG mechanism **shall** be tested for conformance to this Recommendation. A DRBG mechanism **shall** be designed to be tested to ensure that the product is correctly implemented. A testing interface **shall** be available for this purpose in order to allow the insertion of input and the extraction of output for testing.

Implementations to be validated **shall** include the following:

- Documentation specified in Section 11.1.

- Any documentation or results required in derived test requirements.

11.3   Health Testing

A DRBG implementation **shall** perform self-tests to obtain assurance that the DRBG continues to operate as designed and implemented (health testing). The testing function(s) within a DRBG mechanism boundary (or sub-boundary) **shall** test each DRBG mechanism function within that boundary (or sub-boundary), with the possible exception of the test function itself. Note that testing may require the creation and use of one or more instantiations for testing purposes only. A DRBG implementation may optionally perform other self-tests for DRBG functionality in addition to the tests specified in this Recommendation.

All data output from the DRBG mechanism boundary (or sub-boundary) **shall** be inhibited while these tests are performed. The results from known-answer-tests (see Section 11.3.1) **shall not** be output as random bits during normal operation.

11.3.1   Known Answer Testing

Known-answer testing **shall** be conducted as specified below. A known-answer test involves operating the DRBG mechanism with data for which the correct output is already known, and determining if the calculated output equals the expected output (the known answer). The test fails if the calculated output does not equal the known answer. In this case, the DRBG mechanism **shall** enter an error state and output an error indicator (see Section 11.3.6).

Generalized known-answer testing is specified in Sections 11.3.2 to 11.3.5. Testing **shall** be performed on all implemented DRBG mechanism functions, with the possible exception of the test function itself. Documentation **shall** be provided that addresses the continued integrity of the health tests (see Section 11.1).

11.3.2   Testing the Instantiate Function

Known-answer tests on the instantiate function **shall** be performed prior to creating each operational instantiation. However, if several instantiations are performed in quick succession using the same *security_strength* and *prediction_resistance_flag* parameters, then the testing may be reduced to testing only prior to creating the first operational instantiation using that parameter set until such time as the succession of instantiations that use the same *security_strength* and *prediction_resistance_flag* parameters is completed. Thereafter, other instantiations **shall** be tested as specified above.

The *security_strength* and *prediction_resistance_flag* to be used in the operational invocation **shall** be used during the test. Representative fixed values and lengths of the *entropy_input*, *nonce* and *personalization_string* (if supported) **shall** be used; the value of the *entropy_input* used during testing **shall not** be intentionally reused during normal operations (either by the instantiate or the reseed functions). Error handling **shall** also be tested, including whether or not the instantiate function handles an error from the source of entropy input correctly.

If the values used during the test produce the expected results, and errors are handled correctly, then the instantiate function may be used to instantiate using the tested values of *security_strength* and the *prediction_resistance_flag*.

An implementation **should** provide a capability to test the instantiate function on demand.

11.3.3   Testing the Generate Function

Known-answer tests **shall** be performed on the generate function before the first use of the function in an implementation for operational purposes (i.e., the first use ever) and at reasonable intervals defined by the implementer. The implementer **shall** document the intervals and provide a justification for the selected intervals.

The known-answer tests **shall** be performed for each implemented *security_strength*. Representative fixed values and lengths for the *requested_number_of_bits* and *additional_input* (if supported) and the working state of the internal state value (see Sections 8.3 and 10) **shall** be used. If prediction resistance is supported, then each combination of the *security_strength*, *prediction_resistance_request* and *prediction_resistance_flag* **shall** be tested. The error handling for each input parameter **shall** also be tested, and testing **shall** include setting the *reseed_counter* to meet or exceed the *reseed_interval* in order to check that the implementation is reseeded or that the DRBG is uninstantiated, as appropriate (see Section 9.3.1).

If the values used during the test produce the expected results, and errors are handled correctly, then the generate function may be used during normal operations.

Bits generated during health testing **shall not** be output as pseudorandom bits.

An implementation **should** provide a capability to test the generate function on demand.

11.3.4   Testing the Reseed Function

A known-answer test of the reseed function **shall** use the *security_strength* in the internal state of the (testing) instantiation to be reseeded. Representative values of the *entropy_input* and *additional_input* (if supported) and the working state of the internal state value **shall** be used (see Sections 8.3 and 10). Error handling **shall** also be tested, including an error in obtaining the *entropy_input* (e.g., the *entropy_input* source is broken).

If the values used during the test produce the expected results, and errors are handled correctly, then the reseed function may be used to reseed the instantiation.

Self-testing **shall** be performed as follows:

1. When prediction resistance is supported in an implementation, the reseed function **shall** be tested whenever the generate function is tested (see above).

2. When prediction resistance is not supported in an implementation, the reseed function **shall** be tested whenever the reseed function is invoked and before the reseed is performed on the operational instantiation.

An implementation **should** provide a capability to test the reseed function on demand.

11.3.5   Testing the Uninstantiate Function

The uninstantiate function **shall** be tested whenever other functions are tested. Testing **shall** demonstrate that error handling is performed correctly, and the internal state has been zeroized.

11.3.6   Error Handling

The expected errors are indicated for each DRBG mechanism function (see Sections 9.1 - 9.4) and for the derivation functions in Section 10.4. The error handling routines **should** indicate the type of error.

11.3.6.1   Errors Encountered During Normal Operation

Many errors that occur during normal operation may be caused by a consuming application's improper DRBG request; these errors are indicated by "**ERROR_FLAG**" in the pseudocode. In these cases, the consuming application user is responsible for correcting the request within the limits of the user's organizational security policy. For example, if a failure indicating an invalid, requested security strength is returned, a security strength higher than the DRBG or the DRBG instantiation can support has been requested. The user may reduce the requested security strength if the organization's security policy allows the information to be protected using a lower security strength, or the user **shall** use an appropriately instantiated DRBG.

Catastrophic errors (i.e., those errors indicated by the **CATASTROPHIC_ERROR_FLAG** in the pseudocode) detected during normal operation **shall** be treated in the same manner as an error detected during health testing (see Section 11.3.6.2).

11.3.6.2   Errors Encountered During Health Testing

Errors detected during health testing **shall** be perceived as catastrophic DRBG failures.

When a DRBG fails a health test or a catastrophic error is detected during normal operation, the DRBG **shall** enter an error state and output an error indicator. The DRBG **shall not** perform any instantiate, generate or reseed operations while in the error state, and pseudorandom bits **shall not** be output when an error state exists. When in an error state, user intervention (e.g., power cycling of the DRBG) **shall** be required to exit the error state, and the DRBG **shall** be re-instantiated before the DRBG can be used to produce pseudorandom bits. Examples of such errors include:

- A test deliberately inserts an error, and the error is not detected, or

- An incorrect result is returned from the instantiate, reseed or generate function than was expected.

Appendix A: (Normative) Application-Specific Constants

A.1 Constants for the Dual_EC_DRBG

The **Dual_EC_DRBG** requires the specifications of an elliptic curve and two points on the elliptic curve. One of the following NIST **approved** curves with associated points **shall** be used in applications requiring certification under [FIPS 140]. More details about these curves may be found in [FIPS 186]. If alternative points are desired, they **shall** be generated as specified in Appendix A.2.

Each of following curves is given by the equation:

$$y^2 = x^3 - 3x + b \pmod{p}$$

Notation:

p - Order of the field F_p, given in decimal

n - Order of the Elliptic Curve Group, in decimal .

$a - (-3)$ in the above equation

b - Coefficient above

The x and y coordinates of the base point, i.e., generator G, are the same as for the point P.

A.1.1 Curve P-256

p = 115792089210356248762697446949940757353008614\
3415290314195533631308867097853951

n = 115792089210356248762697446949940757352999695\
5224135760342422259061068512044369

b = 5ac635d8 aa3a93e7 b3ebbd55 769886bc 651d06b0 cc53b0f6 3bce3c3e
27d2604b

Px = 6b17d1f2 e12c4247 f8bce6e5 63a440f2 77037d81 2deb33a0
f4a13945 d898c296

Py = 4fe342e2 fe1a7f9b 8ee7eb4a 7c0f9e16 2bce3357 6b315ece
cbb64068 37bf51f5

Qx = c97445f4 5cdef9f0 d3e05e1e 585fc297 235b82b5 be8ff3ef
ca67c598 52018192

Qy = b28ef557 ba31dfcb dd21ac46 e2a91e3c 304f44cb 87058ada
2cb81515 1e610046

A.1.2   Curve P-384

p = 39402006196394479212279040100143613805079739\
        27046544666794829340424572177149687032904726\
        608825893800186160697311231 9

n = 39402006196394479212279040100143613805079739\
        27046544666794690527962765939911326356939895\
        63081522949135544336539426 43

b = b3312fa7 e23ee7e4 988e056b e3f82d19 181d9c6e fe814112 0314088f
        5013875a c656398d 8a2ed19d 2a85c8ed d3ec2aef

Px = aa87ca22 be8b0537 8eb1c71e f320ad74 6e1d3b62 8ba79b98
        59f741e0 82542a38 5502f25d bf55296c 3a545e38 72760ab7

Py = 3617de4a 96262c6f 5d9e98bf 9292dc29 f8f41dbd 289a147c
        e9da3113 b5f0b8c0 0a60b1ce 1d7e819d 7a431d7c 90ea0e5f

Qx = 8e722de3 125bddb0 5580164b fe20b8b4 32216a62 926c5750
        2ceede31 c47816ed d1e89769 124179d0 b6951064 28815065

Qy = 023b1660 dd701d08 39fd45ee c36f9ee7 b32e13b3 15dc0261
        0aa1b636 e346df67 1f790f84 c5e09b05 674dbb7e 45c803dd

A.1.3   Curve P-521

p = 6864797660130609714981900799081393217269435 3\
        00143305409394463459185554318339765605212255 9\
        6406614545549772963113914808580371219879997 1\
        6643812574028291115057151

n = 6864797660130609714981900799081393217269435 3\
        00143305409394463459185554318339765539424505 7\
        7463332171975329639963713633211138647686124 4\
        0380340372808892707005449

b = 051953eb 9618e1c9 a1f929a2 1a0b6854 0eea2da7 25b99b31 5f3b8b48
        9918ef10 9e156193 951ec7e9 37b1652c 0bd3bb1b f073573d f883d2c3
        4f1ef451 fd46b503 f00

Px = c6858e06 b70404e9 cd9e3ecb 662395b4 429c6481 39053fb5
        21f828af 606b4d3d baa14b5e 77efe759 28fe1dc1 27a2ffa8
        de3348b3 c1856a42 9bf97e7e 31c2e5bd 66

Py = 11839296 a789a3bc 0045c8a5 fb42c7d1 bd998f54 449579b4
        46817afb d17273e6 62c97ee7 2995ef42 640c550b 9013fad0
        761353c7 086a272c 24088be9 4769fd16 650

```
Qx = 1b9fa3e5 18d683c6 b6576369 4ac8efba ec6fab44 f2276171
     a4272650 7dd08add 4c3b3f4c 1ebc5b12 22ddba07 7f722943
     b24c3edf a0f85fe2 4d0c8c01 591f0be6 f63

Qy = 1f3bdba5 85295d9a 1110d1df 1f9430ef 8442c501 8976ff34
     37ef91b8 1dc0b813 2c8d5c39 c32d0e00 4a3092b7 d327c0e7
     a4d26d2c 7b69b58f 90666529 11e45777 9de
```

A.2 Using Alternative Points in the Dual_EC_DRBG

The security of **Dual_EC_DRBG** requires that the points P and Q be properly generated. To avoid using potentially weak points, the points specified in Appendix A.1 **should** be used. However, an implementation may use different pairs of points, provided that they are *verifiably random*, as evidenced by the use of the procedure specified in Appendix A.2.1 below, and the self-test procedure in Appendix A.2.2. An implementation that uses alternative points generated by this **approved** method **shall** have them "hard-wired" into its source code, or hardware, as appropriate, and loaded into the *working_state* at instantiation. To conform to this Recommendation, alternatively generated points **shall** use the procedure given in Appendix A.2.1, and verify their generation using Appendix A.2.2.

A.2.1 Generating Alternative P, Q

The curve **shall** be one of the NIST curves from [FIPS 186] that is specified in Appendix A.1 of this Recommendation, and **shall** be appropriate for the desired *security_strength*, as specified in Table 4 in Section 10.3.1.

The points P and Q **shall** be valid base points for the selected elliptic curve that are generated to be verifiably random using the procedure specified in [X9.62]. The following input is required for each point:

An elliptic curve $E = (F_p, a, b)$, cofactor h, prime n, a bitstring *domain_parameter_seed*[7], and hash function **Hash**(). The definition of these parameters is provided in Appendix A.1 of this Recommendation. The *domain_parameter_seed* **shall** be different for each point, and the minimum length m of each *domain_parameter_seed* **shall** conform to Section 10.3.1, Table 4, under "Seed length". The bit length of the *domain_parameter_seed* may be larger than m. The hash function for generating P and Q **shall** be SHA-512 in all cases.

The *domain_parameter_seed* **shall** be different for each point P and Q. A domain parameter seed **shall not** be the seed used to instantiate a DRBG. The domain parameter seed is an arbitrary value that may, for example, be determined from the output of a DRBG.

If the output from the generation procedure in [X9.62] is "failure", a different *domain_parameter_seed* **shall** be used for the point being generated.

---

[7] Called a *SEED* in ANS X9.62.

Otherwise, the output point from the generate procedure in [X9.62] **shall** be used.

A.2.2   Additional Self-testing Required for Alternative *P*, *Q*

To insure that the points *P* and *Q* have been generated appropriately, additional self-test procedures **shall** be performed whenever the instantiate function is invoked. Section 11.3.1 specifies that known-answer tests on the instantiate function be performed prior to creating an operational instantiation. As part of these tests, an implementation of the generation procedure in [X9.62] **shall** be called for each point (i.e., *P* and *Q*) with the appropriate *domain_parameter_seed* value that was used to generate that point. The point returned **shall** be compared with the corresponding stored value of the point. If the generated value does not match the stored value, the implementation **shall** halt with an error condition.

Appendix B: (Normative) Conversion and Auxilliary Routines

B.1 Bitstring to an Integer

Bitstring_to_integer $(b_1, b_2, ..., b_n)$:

    1. $b_1, b_2, ..., b_n$ The bitstring to be converted.

Output:

    1. x The requested integer representation of the bitstring.

Process:

    1. Let $(b_1, b_2, ..., b_n)$ be the bits of b from leftmost to rightmost.

    2. $x = \sum 2^{(n-i)} b_i$.

    3. Return x.

In this Recommendation, the binary length of an integer x is defined as the smallest integer n satisfying $x < 2^n$.

B.2 Integer to a Bitstring

Integer_to_bitstring (x):

    1. x The non-negative integer to be converted.

Output:

    1. $b_1, b_2, ..., b_n$ The bitstring representation of the integer x.

Process:

    1. Let $(b_1, b_2, ..., b_n)$ represent the bitstring, where $b_1 = 0$ or 1, and b_1 is the most significant bit, while b_n is the least significant bit.

    2. For any integer n that satisfies $x < 2^n$, the bits b_i **shall** satisfy:

$$x = \sum 2^{(n-i)} b_i.$$

    3. Return $b_1, b_2, ..., b_n$.

In this Recommendation, the binary length of the integer x is defined as the smallest integer n that satisfies $x < 2^n$.

B.3 Integer to an Byte String

Integer_to_byte_string (x):

1. A non-negative integer x, and the intended length n of the byte string satisfying

$$2^{8n} > x.$$

Output:

1. A byte string O of length n bytes.

Process:

1. Let $O_1, O_2, ..., O_n$ be the bytes of O from leftmost to rightmost.

2. The bytes of O **shall** satisfy:

$$x = \Sigma\, 2^{8(n-i)} O_i$$

for $i = 1$ to n.

3. Return O.

B.4   Byte String to an Integer

Byte_string_to_integer (O):

1. A byte string O of length n bytes.

Output:

1. A non-negative integer x.

Process:

1. Let $O_1, O_2, ..., O_n$ be the bytes of O from leftmost to rightmost.

2. x is defined as follows:

$$x = \Sigma\, 2^{8(n-i)} O_i$$

for $i = 1$ to n.

3. Return x.

B.5   Converting Random Numbers from/to Random Bits

The random values required for cryptographic applications are generally of two types: either a random bitstring of a specified length, or a random integer in a specified interval. In some cases, a DRBG may return a random number in a specified interval that needs to be converted into random bits; in other cases, a DRBG returns a random bitstring that needs to be converted to a random number in a specific range.

B.5.1   Converting Random Bits into a Random Number

In some cryptographic applications sequences of random numbers are required (a_0, a_1, a_2,...) where:

i) Each integer a_i satisfies $0 \leq a_i \leq r\text{-}1$, for some positive integer r (the *range* of the random numbers);

ii) The equation $a_i = s$ holds, with probability almost exactly $1/r$, for any $i \geq 0$ and for any s ($0 \leq s \leq r\text{-}1$);

iii) Each value a_i is statistically independent of any set of values a_j ($j \neq i$).

Four techniques are specified for generating sequences of random numbers from sequences of random bits.

If the range of the number required is $a \leq a_i \leq b$ rather than $0 \leq a_i \leq r\text{-}1$, then a random number in the desired range can be obtained by computing $a_i + a$, where a_i is a random number in the range $0 \leq a_i \leq b\text{-}a$ (that is, when $r = b\text{-}a\text{+}1$).

B.5.1.1 The Simple Discard Method

Let m be the number of bits needed to represent the value $(r\text{–}1)$. The following method may be used to generate the random number a:

1. Use the random bit generator to generate a sequence of m random bits, $(b_0, b_1, \ldots, b_{m\text{-}1})$.

2. Let $c = \sum_{i=0}^{m-1} 2^i b_i$.

3. If $c < r$ then put $a = c$, else discard c and go to Step 1.

This method produces a random number a with no skew (no bias). A possible disadvantage of this method, in general, is that the time needed to generate such a random a is not a fixed length of time because of the conditional loop.

The ratio $r/2^m$ is a measure of the efficiency of the technique, and this ratio will always satisfy $0.5 < r/2^m \leq 1$. If $r/2^m$ is close to 1, then the above method is simple and efficient. However, if $r/2^m$ is close to 0.5, then the simple discard method is less efficient, and the more complex method below is recommended.

B.5.1.2 The Complex Discard Method

Choose a small positive integer t (the number of same-size random number outputs desired), and then let m be the number of bits in $(r^t \text{–}1)$. This method may be used to generate a sequence of t random numbers $(a_0, a_1, \ldots, a_{t\text{-}1})$:

1. Use the random bit generator to generate a sequence of m random bits, $(b_0, b_1, \ldots, b_{m\text{-}1})$.

2. Let $c = \sum_{i=0}^{m-1} 2^i b_i$.

3. If $c < r^t$, then

let $(a_0, a_1, \ldots, a_{t-1})$ be the unique sequence of values satisfying $0 \leq a_i \leq r-1$ such that $c = \sum_{i=0}^{t-1} r^i a_i$.

else discard c and go to Step 1.

This method produces random numbers $(a_0, a_1, \ldots, a_{t-1})$ with no skew. A possible disadvantage of this method, in general, is that the time needed to generate these numbers is not a fixed length of time because of the conditional loop. The complex discard method may have better overall performance than the simple discard method if many random numbers are needed.

The ratio $r^t/2^m$ is a measure of the efficiency of the technique, and this ratio will always satisfy $0.5 < r^t/2^m \leq 1$. Hence, given r, it is recommended to choose t so that t is the smallest value such that $r^t/2^m$ is close to 1. For example, if $r = 3$, then choosing $t = 3$ means that $m = 5$ (as r^t is 27) and $r^t/m = 27/32 \approx 0.84$, and choosing $t = 5$ means that $m = 8$ (as r^t is 243) and $r^t/m = 243/256 \approx 0.95$. The complex discard method coincides with the simple discard method when $t = 1$.

B.5.1.3 The Simple Modular Method

Let m be the number of bits needed to represent the value $(r-1)$, and let s be a security parameter. The following method may be used to generate one random number a:

1. Use the random bit generator to generate a sequence of $m+s$ random bits, $(b_0, b_1, \ldots, b_{m+s-1})$.

2. Let $c = \sum_{i=0}^{m+s-1} 2^i b_i$.

3. Let $a = c \bmod r$.

The simple modular method can be coded to take constant time. This method produces a random value with a negligible skew, that is, the probability that $a_i = w$ for any particular value of w ($0 \leq w \leq r-1$) is not exactly $1/r$. However, for a large enough value of s, the difference between the probability that $a_i = w$ for any particular value of w and $1/r$ is negligible. The value of s **shall** be greater than or equal to 64.

B.5.1.4 The Complex Modular Method

Choose a small positive integer t (the number of same-size random number outputs desired) and a security parameter s; let m be the number of bits in $(r^t - 1)$. The following method may be used to generate a sequence of t random numbers $(a_0, a_1, \ldots, a_{t-1})$:

1. Use the random bit generator to generate a sequence of $m+s$ random bits, $(b_0, b_1, \ldots, b_{m+s-1})$.

2. Let $\sum_{i=0}^{m+s-1} 2^i b_i \bmod r^t$.

3. Let $(a_0, a_1, \ldots, a_{t-1})$ be the unique sequence of values satisfying $0 \leq a_i \leq r\text{-}1$ such that $c = \sum_{i=0}^{t-1} r^i a_i$.

The complex modular method may have better overall performance than the simple modular method if many random numbers are needed. This method produces a random value with a negligible skew; that is, the probability that $a_i = w$ for any particular value of w ($0 \leq w \leq r\text{-}1$) is not exactly $1/r$. However, for a large enough value of s, the difference between the probability that $a_i = w$ for any particular value of w and $1/r$ is negligible. The value of s **shall** be greater than or equal to 64. The complex modular method coincides with the simple modular method when $t = 1$.

B.5.2   Converting a Random Number into Random Bits

B.5.2.1 The No Skew (Variable Length Extraction) Method

This is a method of extracting random unbiased bits from a random number modulo a number n. First, a toy example is provided in order to explain how the method works, and then pseudocode is given.

For the toy example, the insight is to look at the modulus n and the random number r as bits, from left to right, and to partition the possible values of r into disjoint sets based on the largest number of random bits that might be extracted. As a small example, if $n = 11$, then the binary representation of n is b'1011', and the possible values of r (in binary) are as follows:

0000, 0001, 0010, 0011, 0100, 0101, 0110, 0111, 1000, 1001, 1010.

Let the leftmost bit be considered as the bit 4, and the rightmost bit be considered as the bit 1.

1. As the 4th bit of n is b'1', look at the 4th bit of r.

2. If the 4th bit of r is b'0', then the remaining 3 bits can be extracted as unbiased random bits. This forms a class of [0000, 0001, 0010, 0011, 0100, 0101, 0110, 0111] and maps each respective element into the 3-bit sequences [000, 001, 010, 011, 100, 0101, 110, 111], each of which is unbiased, and the process is completed

3. If the 4th bit of r is b'1', then r falls into the remainder [1000, 1001, 1010], and the process needs to continue with step 4 in order to extract unbiased bits.

4. As the 3rd bit of n is b'0', the 3rd bit of r is always b'0' in the class determined in step 3; therefore the 3rd bit of r is already known to be biased, so the analysis moves to the next bit (step 5).

5. The 2nd bit of n is b'1', so this forms a subclass [1000, 1001], from which one random unbiased bit can be extracted, namely the 1st bit.

The remaining value of 1010 cannot be used to extract random bits. However, obtaining this value is not usual. For this tiny example: 8/11 of the time, 3 unbiased random bits can be extracted; 2/11 of the time, 1 unbiased bit can be extracted; and

1/11, no unbiased bits can be extracted. As can be seen, it is not known ahead of time how many unbiased bits will be able to be extracted, although the average will be known.

Let both the modulus n and the random r values have m bits. This means that the mth bit of $n = $ b'1', although m^{th} bit of r may be either b'1' or b'0'.

1. $j = 0$.

2.   Do $i = m$ to 1 by -1

> Comment: if the ith bit of $n = $ b'0', or the ith bit of $r = $ b'1', then this is a skew situation; the routine cannot extract i-1 unbiased bits, so the index is shifted right to check next bit

    2.1   If (((the ith bit of $n = $ b'0') or (the ith bit of $r = $ b'1')), then go to step 2.5.

    2.2 $j = i$-1.

    2.3   *output* = the jth bit of r.

    2.4 $i = 1$ Comment: all unbiased bits possible have been extracted, so exit .

    2.5   Continue

The extraction takes a variable amount of time, but this varying amount of time does not leak any information to a potential adversary that can be used to attack the method.

B.5.2.2   The Negligible Skew (Fixed Length Extraction) Method

A possible disadvantage of the No Skew (Variable Length Extraction) Method of Appendix B.5.2.1 is that it takes a variable amount of time to extract a variable number of random bits. To address this concern and to simplify the extraction method, the following method is specified that extracts a fixed number of random bits with a negligible skew. This method exploits the fact that the modulus n is known before the extraction occurs.

1. Examine the modulus considered as a binary number from left to right, and determine the index bit such that there are at least 16 b'1' bits to the left. Call this bit i.

2. Extract random bits from the random number r by truncating on the left up to bit i. This is the output = $r(i,1)$.

This method is especially appropriate when the high order bits of the modulus are all set to b'1' for efficiency reasons, as is the case with the NIST elliptic curves over prime fields.

This method is acceptable for elliptic curves, based on the following analysis. When considering the no skew method, once the random bits are extracted, it is obvious that less than the full number of random bits can be extracted, and the extraction result will still be random. The truncation of more bits than necessary is acceptable. What about truncation of too few bits? For a random number, the no skew extraction process would continue only if the 16 bits of r corresponding to the b'1' bits in n are all zero. For a random number, this occurs about once every 2^{16} times. As the modulus is at least 160 bits, this means that 144 bits with a skew are extracted in this case. On average, once every 9,437,184 output bits (or more), there will be a 144-bit substring somewhere in that total that has a skew, which will have the leftmost bit or bits tending to a binary zero bit or bits. This skew could be as little as one bit. However, an adversary will not know exactly where this skewed substring occurs. The 9,437,184 total output bits will still be overwhelmingly likely to be within the statistical variation of a random bitstring; that is, the statistical variation almost certainly will be much greater than this negligible skew.

Appendix C: (Informative) Security Considerations when Extracting Bits in the Dual_EC_DRBG

C.1 Potential Bias Due to Modular Arithmetic for Curves Over F_p

Given an integer x in the range 0 to 2^N-1, where N is any positive integer, the rth bit of x depends solely upon whether $\left\lfloor \dfrac{x}{2^r} \right\rfloor$ is odd or even. Exactly ½ of the integers in this range have the property that their r^{th} bit is 0. But if x is restricted to F_P, i.e., to the range 0 to p-1, this statement is no longer true.

By excluding the $k = 2^N - p$ values $p, p+1, ..., 2^N -1$ from the set of all integers in Z_{2^N}, the ratio of ones to zeroes in the rth bit is altered from $2^{N-1} / 2^{N-1}$ to a value that can be no smaller than $(2^{N-1} - k)/ 2^{N-1}$. For all the primes p used in this Recommendation, $k/2^{N-1}$ is smaller than 2^{-31}. Thus, the ratio of ones to zeroes in any bit is within at least 2^{-31} of 1.0.

To detect this small difference from random, a sample of at least 2^{64} outputs is required before the observed distribution of 1's and 0's is more than one standard deviation away from flat random. This effect is dominated by the bias addressed below in Appendix C.2.

C.2 Adjusting for the Missing Bit(s) of Entropy in the x Coordinates.

In a truly random sequence, it should not be possible to predict any bits from previously observed bits. With the **Dual_EC_DRBG,** the full output block of bits produced by the algorithm is "missing" some entropy. Fortunately, by discarding some of the bits, those bits remaining can be made to have nearly "full strength", in the sense that the entropy that they are missing is negligibly small.

To illustrate what can happen, suppose that the curve P-256 is selected, and that all 256 bits produced were output by the generator, i.e. that *outlen* = 256 also. Suppose also that 255 of these bits are published, and the 256-th bit is kept "secret". About ½ the time, the unpublished bit could easily be determined from the other 255 bits. Similarly, if 254 of the bits are published, about ¼ of the time the other two bits could be predicted. This is a simple consequence of the fact that only about 1/2 of all 2^m bitstrings of length m occur in the list of all x coordinates of curve points.

The "abouts" in the preceding example can be made more precise, taking into account the difference between 2^m and p, and the actual number of points on the curve (which is always within $2 * p^{\frac{1}{2}}$ of p). For the curves in this Recommendation, these differences do not matter at the scale of the results, so they will be ignored. This allows the heuristics given here to work for any curve with "about" $(2^m)/f$ points, where $f = 1$ is the curve's cofactor. For all the curves in this Recommendation, the cofactor $f = 1$.

The basic assumption needed is that the approximately $(2^m)/(2f)$ x coordinates that do occur are "uniformly distributed": a randomly selected m-bit pattern has a probability $1/2f$ of being an x coordinate. The assumption allows a straightforward calculation, albeit

approximate, for the entropy in the rightmost (least significant) m-d bits of
Dual_EC_DRBG output, with $d \ll m$.

The formula is $\quad E = -\sum_{j=0}^{2^d}\left[2^{m-d}\, binomprob\!\left(2^d\,.z.2^d - j\right)\right]p_j\,\log_2 p_j$, where E is the entropy.

For each $0 \le j \le 2^d$, the term in braces represents the approximate number of bitstrings b of length $(m$-$d)$ such that there are exactly j points whose x-coordinates have their $(m$-$d)$ least significant bits equal to b; $z = (2f$-$1)/2f$ is the probability that any particular string occurs in an x coordinate; $p_j = (j*2f)/2^m$ is the probability that a member of the j-th category occurs. Note that the j=0 category contributes nothing to the entropy (randomness).

The values of E for d up to 16 are:

$\log 2(f)$: 0 d: 0   entropy: 255.00000000 m-d: 256

$\log 2(f)$: 0 d: 1   entropy: 254.50000000 m-d: 255

$\log 2(f)$: 0 d: 2   entropy: 253.78063906 m-d: 254

$\log 2(f)$: 0 d: 3   entropy: 252.90244224 m-d: 253

$\log 2(f)$: 0 d: 4   entropy: 251.95336161 m-d: 252

$\log 2(f)$: 0 d: 5   entropy: 250.97708960 m-d: 251

$\log 2(f)$: 0 d: 6   entropy: 249.98863897 m-d: 250

$\log 2(f)$: 0 d: 7   entropy: 248.99434222 m-d: 249

$\log 2(f)$: 0 d: 8   entropy: 247.99717670 m-d: 248

$\log 2(f)$: 0 d: 9   entropy: 246.99858974 m-d: 247

$\log 2(f)$: 0 d: 10   entropy: 245.99929521 m-d: 246

$\log 2(f)$: 0 d: 11   entropy: 244.99964769 m-d: 245

$\log 2(f)$: 0 d: 12   entropy: 243.99982387 m-d: 244

$\log 2(f)$: 0 d: 13   entropy: 242.99991194 m-d: 243

$\log 2(f)$: 0 d: 14   entropy: 241.99995597 m-d: 242

$\log 2(f)$: 0 d: 15   entropy: 240.99997800 m-d: 241

$\log 2(f)$: 0 d: 16   entropy: 239.99998900 m-d: 240

The analysis above uses Shannon entropy. However for this Recommendation, min-entropy is a more appropriate measure of randomness than Shannon entropy. If the analysis above is repeated for min-entropy, then one finds that about one bit of min-entropy is missing for most values of $d < m/2$. The main reason for this is that the case of $j = 2^d$ is expected to occur, provided that $d < m/2$. Therefore, the maximum probability for a particular bit string of length m-d is $p_{2^d} = 2^{d+1-m}$, which gives a min-entropy of m-d-1. An adversary attempting to guess the value of the bitstring of length m-d, would choose a string such that $j = 2^d$. On the other hand, generally speaking, the security strength associated with an m-bit elliptic curve is only $m/2$ bits, which implies that only $m/2$ bits of min-entropy are required. Therefore, the loss of a single bit of min-entropy may be deemed acceptable here because the min-entropy would still be well over what is needed.

Observations:

  a) The table starts where it should, at one missing bit;

  b) The missing entropy rapidly decreases;

  c) For the curves in this Recommendation, d=13 leaves one bit of information in every 10,000 (m-13)-bit outputs (i.e., one bit of entropy is missing in a collection of 10,000 outputs).

Based on these calculations, for the mod p curves, it is recommended that an implementation **shall** remove at least the leftmost (most significant) 13 bits of every m-bit output.

For ease of implementation, the value of d **should** be adjusted upward, if necessary, until the number of bits remaining, m-d= $outlen$, is a multiple of eight. By this rule, the recommended number of bits discarded from each x-coordinate will be either 16 or 17. As noted in Section 10.3.1.4, an implementation may truncate additional bits from each x-coordinate, provided that the number retained is a multiple of eight.

Because only half of all values in [0, 1, ..., p-1] are valid x-coordinates on an elliptic curve defined over F_p, it is clear that full x-coordinates **should not** be used as pseudorandom bits. The solution to this problem is to truncate these x-coordinates by removing the high order 16 or 17 bits. The entropy loss associated with such truncation amounts has been demonstrated to be minimal (see the above chart).

One might wonder if it would be desirable to truncate more than this amount. The obvious drawback to such an approach is that increasing the truncation amount hinders the performance. However, there is an additional reason that argues against increasing the truncation. Consider the case where the low s bits of each x-coordinate are kept. Given some subinterval I of length 2^s contained in [0, p), and letting $N(I)$ denote the number of x-coordinates in I, recent results on the distribution of x-coordinates in [0, p) provide the following bound:

$$\left| \frac{N(I)}{(p/2)} - \frac{2^s}{p} \right| < \frac{k * \log^2 p}{\sqrt{p}},$$

where k is some constant derived from the asymptotic estimates given in [Shparlinski]. For the case of P-521, this is roughly equivalent to:

$$\left| N(I) - 2^{(s-1)} \right| < k * 2^{277},$$

where the constant k is independent of the value of s. For $s < 2^{277}$, this inequality is weak and provides very little support for the notion that these truncated x-coordinates are uniformly distributed. On the other hand, the larger the value of s, the sharper this inequality becomes, providing stronger evidence that the associated truncated x-coordinates are uniformly distributed. Therefore, by keeping truncation to an acceptable minimum, the performance is increased, and certain guarantees can be made about the uniform distribution of the resulting truncated quantities. Further discussion of the uniformity of the truncated x-coordinates is found in [Gurel], where the form of the prime defining the field is also taken into account.

Appendix D: (Informative) Example Pseudocode for Each DRBG Mechanism

The internal states in these examples are considered to be an array of states, identified by *state_handle*. A particular state is addressed as *internal_state* (*state_handle*), where the value of *state_handle* begins at 0 and ends at *n*-1, and *n* is the number of internal states provided by an implementation. A particular element in the internal state is addressed by *internal_state* (*state_handle*).*element*. In an empty internal state, all bitstrings are set to *Null*, and all integers are set to 0.

For each example in this appendix, arbitary values have been selected that are consistent with the allowed values for each DRBG mechanism, as specified in the appropriate table in Section 10.

The pseudocode in this appendix does not include the necessary conversions (e.g., integer to bitstring) for an implementation. When conversions are required, they **shall** be accomplished as specified in Appendix B.

The following routine is defined for these pseudocode examples:

Find_state_space (): A function that finds an unused internal state. The function returns a *status* (either "Success" or a message indicating that an unused internal state is not available) and, if *status* = "Success", a *state_handle* that points to an available *internal_state* in the array of internal states. If *status* ≠ "Success", an invalid *state_handle* is returned.

When the uninstantantiate function is invoked in the following examples, the function specified in Section 9.4 is called.

D.1 Hash_DRBG Example

This example of **Hash_DRBG** uses the SHA-1 hash function, and prediction resistance is supported. Both a personalization string and additional input are supported. A 32-bit incrementing counter is used as the nonce for instantiation (*instantiation_nonce*); the nonce is initialized when the DRBG is instantiated (e.g., by a call to the clock or by setting it to a fixed value) and is incremented for each instantiation.

A total of ten internal states are provided (i.e., ten instantiations may be handled simultaneously).

For this implementation, the functions and algorithms are "inline", i.e., the algorithms are not called as separate routines from the function envelopes. Also, the **Get_entropy_input** function uses only three input parameters, since the first two parameters (as specified in Section 9) have the same value.

The internal state contains values for *V*, *C*, *reseed_counter*, *security_strength* and *prediction_resistance_flag*, where *V* and *C* are bitstrings, and *reseed_counter*, *security_strength* and the *prediction_resistance_flag* are integers. A requested prediction resistance capability is indicated when *prediction_resistance_flag* = 1.

In accordance with Table 2 in Section 10.1, the 112- and 128-bit security strengths may be instantiated. Using SHA-1, the following definitions are applicable for the instantiate, generate and reseed functions and algorithms:

1. *highest_supported_security_strength* = 128.

2. Output block length (*outlen*) = 160 bits.

3. Required minimum entropy for instantiation and reseed = *security_strength*.

4. Seed length (*seedlen*) = 440 bits.

5. Maximum number of bits per request (*max_number_of_bits_per_request*) = 5000 bits.

6. Reseed interval (*reseed_interval*) = 100,000 requests.

7. Maximum length of the personalization string (*max_personalization_string_length*) = 512 bits.

8. Maximum length of additional_input (*max_additional_input_string_length*) = 512 bits.

9. Maximum length of entropy input (*max _length*) = 1000 bits.

D.1.1   Instantiation of Hash_DRBG

This implementation will return a text message and an invalid state handle (−1) when an error is encountered. Note that the value of *instantiation_nonce* is an internal value that is always available to the instantiate function.

Note that this implementation does not check the *prediction_resistance_flag*, since the implementation has been designed to support prediction resistance. However, if a consuming application actually wants prediction resistance, the implementation expects that *prediction_resistance_flag* = 1 during instantiation; this will be used in the generate function in Appendix D.1.3.

Hash_DRBG_Instantiate_function:

> **Input:** integer (*requested_instantiation_security_strength, prediction_resistance_flag*), bitstring *personalization_string*.

> **Output:** string *status*, integer *state_handle*.

> **Process:**

> > Comment. Check the input parameters.

> 1. If (*requested_instantiation_security_strength* > 128), then **Return** ("Invalid *requested_instantiation_security_strength*", −1).

> 2. If (**len** (*personalization_string*) > 512), then **Return** ("*Personalization_string* too long", −1).

> Comment: Set the *security_strength* to one of the valid security strengths.

3. If (*requested_instantiation_security_strength* $\leq$ 112), then *security_strength* = 112

    Else *security_strength* = 128.

> Comment: Get the *entropy_input*.

4. (*status*, *entropy_input*) = **Get_entropy_input** (*security_strength*, 1000, *prediction_resistance_request*).

5. If (*status* $\neq$ "Success"), then **Return** ("Catastrophic failure of the *entropy_input* source:" $\|$ *status*, –1).

> Comment: Increment the nonce; actual coding must ensure that it wraps when the storage limit is reached.

6. *instantiation_nonce* = *instantiation_nonce* + 1.

> Comment: The instantiate algorithm is provided in steps 7-11.

7. *seed_material* = *entropy_input* $\|$ *instantiation_nonce* $\|$ *personalization_string*.

8. *seed* = **Hash_df** (*seed_material*, 440).

9. *V* = *seed*.

10. *C* = **Hash_df** ((0x00 $\|$ *V*), 440).

11. *reseed_counter* = 1.

> Comment: Find an unused internal state.

12. (*status*, *state_handle*) = **Find_state_space** ( ).

13. If (*status* $\neq$ "Success"), then **Return** (*status*, –1).

14. Save the internal state.

    14.1 *internal_state* (*state_handle*).*V* = *V*.

    14.2 *internal_state* (*state_handle*).*C* = *C*.

    14.3 *internal_state* (*state_handle*).*reseed_counter* = *reseed_counter*.

    14.4 *internal_state* (*state_handle*). *security_strength* = *security_strength*.

    14.5 *internal_state* (*state_handle*).*prediction_resistance_flag* = *prediction_resistance_flag*.

15. **Return** ("Success", *state_handle*).

D.1.2   Reseeding a Hash_DRBG Instantiation

The implementation is designed to return a text message as the *status* when an error is encountered.

Hash_DRBG_Reseed_function:

Input: integer *state_handle,* integer *prediction_resistance_request,* bitstring *additional_input.*

Output: string *status.*

Process:

> Comment: Check the validity of the *state_handle.*

1. If ((*state_handle* < 0) or (*state_handle* > 9) or (*internal_state* (*state_handle*) = {*Null, Null,* 0, 0, 0})), then **Return** ("State not available for the *state_handle*").

> Comment: Get the internal state values needed to determine the new internal state.

2. Get the appropriate *internal_state* values.

> $V = internal\_state(state\_handle).V.$
>
> $security\_strength = internal\_state(state\_handle).security\_strength.$
>
> Check the length of the *additional_input.*

3. If (**len** (*additional_input*) > 512), then **Return** ("*additional_input* too long").

> Comment: Get the *entropy_input.*

4. (*status, entropy_input*) = **Get_entropy_input** (*security_strength,* 1000, *prediction_resistance_request*).

5. If (*status* ≠ "Success"), then **Return** ("Catastrophic failure of the *entropy_input* source:" || *status*).

> Comment: The reseed algorithm is provided in steps 6-10.

6. *seed_material* = 0x01 || *V* || *entropy_input* || *additional_input.*

7. *seed* = **Hash_df** (*seed_material,* 440).

8. *V* = *seed.*

9. *C* = **Hash_df** ((0x00 || *V*), 440).

10. *reseed_counter* = 1.

> Comment: Update the *working_state* portion of the internal state.

11. Update the appropriate *state* values.

11.1 *internal_state (state_handle).V = V.*

11.2 *internal_ state (state_handle).C = C.*

11.3 *internal_ state (state_handle).reseed_counter = reseed_counter.*

12. **Return** ("Success").

D.1.3   Generating Pseudorandom Bits Using Hash_DRBG

The implementation returns a *Null* string as the pseudorandom bits if an error has been detected. Prediction resistance is requested when *prediction_resistance_request* = 1.

In this implementation, prediction resistance is requested by supplying *prediction_resistance_request* = 1 when the **Hash_DRBG** function is invoked.

Hash_DRBG_Generate_function:

> **Input:** integer (*state_handle, requested_no_of bits, requested_security_strength, prediction_resistance_request*), bitstring *additional_input*.

> **Output:** string *status*, bitstring *pseudorandom_bits*.

> **Process:**

> Comment: Check the validity of the *state_handle*.

1. If ((*state_handle* < 0) or (*state_handle* > 9) or (*state (state_handle)* = {*Null, Null*, 0, 0, 0})), then **Return** ("State not available for the *state_handle*", *Null*).

2. Get the internal state values.

   2.1 *V = internal_state (state_handle).V.*

   2.2 *C = internal_state (state_handle).C.*

   2.3 *reseed_counter = internal_state (state_handle).reseed_counter.*

   2.4 *security_strength = internal_state (state_handle).security_strength.*

   2.5 *prediction_resistance_flag = internal_state (state_handle).prediction_resistance_flag.*

   > Comment: Check the validity of the other input parameters.

3. If (*requested_no_of_bits* > 5000) then **Return** ("Too many bits requested", *Null*).

4. If (*requested_security_strength* > *security_strength*), then **Return** ("Invalid *requested_security_strength*", *Null*).

5. If (**len** (*additional_input*) > 512), then **Return** ("*additional_input* too long", *Null*).

6. If ((*reseed_counter* > 100,000) OR (*prediction_resistance_request* = 1)), then

6.1   *status* = **Hash_DRBG_Reseed_ function** (*state_handle,*
prediction_resistance_request, additional_input).

6.2   If (*status* ≠ "Success"), then **Return** (*status, Null*).

6.3   Get the new internal state values that have changed.

7.3.1 *V* = *internal_state* (*state_handle*).*V*.

7.3.2 *C* = *internal_state* (*state_handle*).*C*.

7.3.3 *reseed_counter* = *internal_state* (*state_handle*).*reseed_counter*.

6.4   *additional_input* = *Null*.

Comment: Steps 7-15 provide the rest of the generate algorithm. Note that in this implementation, the **Hashgen** routine is also inline as steps 8-12.

7. If (*additional_input* ≠ *Null*), then do

7.1 *w* = **Hash** (0x02 || *V* || *additional_input*).

7.2 *V* = (*V* + *w*) mod 2^{440}.

8. $m = \left\lceil \dfrac{requested\_no\_of\_bits}{outlen} \right\rceil$.

9.   *data* = *V*.

10. *W* = the Null string.

11. For *i* = 1 to *m*

11.1 w_i = **Hash** (*data*).

11.2 *W* = *W* || w_i.

11.3 *data* = (*data* + 1) mod 2^{440}.

12. *pseudorandom_bits* = Leftmost (*requested_no_of_bits*) bits of *W*.

13. *H* = **Hash** (0x03 || *V*).

14. *V* = (*V* + *H* + *C* + *reseed_counter*) mod 2^{440}.

15. *reseed_counter* = *reseed_counter* + 1.

Comments: Update the *working_state*.

16. Update the changed values in the *state*.

16.1   *internal_state* (*state_handle*).*V* = *V*.

16.2   *internal_state* (*state_handle*).*reseed_counter* = *reseed_counter*.

17. **Return** ("Success", *pseudorandom_bits*).

D.2 HMAC_DRBG Example

This example of **HMAC_DRBG** uses the SHA-256 hash function. Reseeding and prediction resistance are not supported. The nonce for instantiation consists of a random value with *security_strength*/2 bits of entropy; the nonce is obtained by increasing the call for entropy bits via the **Get_entropy_input** call by *security_strength*/2 bits (i.e., by adding *security_strength*/2 bits to the *security_strength* value). The **HMAC_DRBG_Update** function is specified in Section 10.1.2.2.

A personalization string is supported, but additional input is not. A total of three internal states are provided. For this implementation, the functions and algorithms are written as separate routines. Also, the **Get_entropy_input** function uses only two input parameters, since the first two parameters (as specified in Section 9) have the same value, and prediction resistance is not available.

The internal state contains the values for *V*, *Key*, *reseed_counter*, and *security_strength*, where *V* and *C* are bitstrings, and *reseed_counter* and *security_strength* are integers.

In accordance with Table 2 in Section 10.1, security strengths of 112, 128, 192 and 256 bits may be instantiated. Using SHA-256, the following definitions are applicable for the instantiate and generate functions and algorithms:

1. *highest_supported_security_strength* = 256.

2. Output block (*outlen*) = 256 bits.

3. Required minimum entropy for the entropy input at instantiation = 3/2 *security_strength* (this includes the entropy required for the nonce).

4. Seed length (*seedlen*) = 440 bits.

5. Maximum number of bits per request (*max_number_of_bits_per_request*) = 7500 bits.

6. Reseed_interval (*reseed_ interval*) = 10,000 requests.

7. Maximum length of the personalization string (*max_personalization_string_length*) = 160 bits.

8. Maximum length of the entropy input (*max _length*) = 1000 bits.

D.2.1   Instantiation of HMAC_DRBG

This implementation will return a text message and an invalid state handle (−1) when an error is encountered.

HMAC_DRBG_Instantiate_function:

>**Input:** integer (*requested_instantiation_security_strength*), bitstring
>   *personalization_string*.

>**Output:** string *status*, integer *state_handle*.

>**Process:**

Check the validity of the input parameters.

1. If (*requested_instantiation_security_strength* > 256), then **Return** ("Invalid *requested_instantiation_security_strength*", –1).

2. If (**len** (*personalization_string*) > 160), then **Return** ("*Personalization_string* too long*", –1)

> Comment: Set the *security_strength* to one of the valid security strengths.

3. If (*requested_security_strength* ≤ 112), then *security_strength* = 112

   Else (*requested_ security_strength* ≤ 128), then *security_strength* = 128

   Else (*requested_ security_strength* ≤ 192), then *security_strength* = 192

   Else *security_strength* = 256.

> Comment: Get the *entropy_input and the nonce*.

4. *min_entropy* = 1.5 × *security_strength*.

5. (*status, entropy_input*) = **Get_entropy_input** (*min_entropy*, 1000).

6. If (*status* ≠ "Success"), then **Return** ("Catastrophic failure of the entropy source:" || *status*, –1).

> Comment: Invoke the instantiate algorithm. Note that the *entropy_input* contains the nonce.

7. (*V, Key, reseed_counter*) = **HMAC_DRBG_Instantiate_algorithm** (*entropy_input, personalization_string*).

> Comment: Find an unused internal state.

8. (*status, state_handle*) = **Find_state_space** ( ).

9. If (*status* ≠ "Success"), then **Return** ("No available state space:" || *status*, –1).

10. Save the initial state.

    10.1 *internal_state (state_handle).V = V.*

    10.2 *internal state (state_handle). Key = Key.*

    10.3 *internal state (state_handle). reseed_counter = reseed_counter.*

    10.4 *internal_state (state_handle).security_strength = security_strength.*

11. Return ("Success" and *state_handle*).

HMAC_DRBG_Instantiate_algorithm (...):

   Input: bitstring (*entropy_input, personalization_string*).

   Output: bitstring (*V, Key*), integer *reseed_counter*.

Process:

1. *seed_material = entropy_input || personalization_string*.

2. Set *Key* to *outlen* bits of zeros.

3. Set *V* to *outlen*/8 bytes of 0x01.

4. (*Key, V*) = **HMAC_DRBG_Update** (*seed_material, Key, V*).

5. *reseed_counter* = 1.

6. **Return** (*V, Key, reseed_counter*).

D.2.2   Generating Pseudorandom Bits Using HMAC_DRBG

The implementation returns a *Null* string as the pseudorandom bits if an error has been detected.

HMAC_DRBG_Generate_function:

Input: integer (*state_handle, requested_no_of_bits, requested_security_strength*).

Output: string (*status*), bitstring *pseudorandom_bits*.

Process:

> Comment: Check for a valid state handle.

1. If ((*state_handle* < 0) or (*state_handle* > 2) or (*internal_state (state_handle*) = {*Null, Null*, 0, 0}), then **Return** ("State not available for the indicated *state_handle*", *Null*).

2. Get the internal state.

   2.1 *V = internal_state (state_handle).V*.

   2.2 *Key = internal_state (state_handle).Key*.

   2.3 *security_strength = internal_state (state_handle).security_strength*.

   2.4 *reseed_counter = internal_state (state_handle).reseed_counter*.

> Comment: Check the validity of the rest of the input parameters.

3. If (*requested_no_of_bits* > 7500), then **Return** ("Too many bits requested", *Null*).

4. If (*requested_security_strength > security_strength*), then **Return** ("Invalid *requested_security_strength*", *Null*).

> Comment: Invoke the generate algorithm.

5. (*status, pseudorandom_bits, V, Key, reseed_counter*) = **HMAC_DRBG_Generate_algorithm** (*V, Key, reseed_counter, requested_number_of_bits*).

6. If (*status* = "Reseed required"), then **Return** ("DRBG can no longer be used. A new instantiation is required", *Null*).

7. Update the changed state values.

   7.1 *internal_state (state_handle).V = V.*

   7.2 *internal_state (state_handle).Key = Key.*

   7.3 *internal_state (state_handle).reseed_counter = reseed_counter.*

8. **Return** ("Success", *pseudorandom_bits*).

HMAC_DRBG_Generate_algorithm:

Input: bitstring (*V, Key*), integer (*reseed_counter, requested_number_of_bits*).

Output: string *status*, bitstring (*pseudorandom_bits, V, Key*), integer *reseed_counter*.

Process:

1 If (*reseed_counter* ≥ 10,000), then **Return** ("Reseed required", *Null, V, Key, reseed_counter*).

2. *temp = Null.*

3 While (**len** (*temp*) < *requested_no_of_bits*) do:

   3.1 *V* = **HMAC** (*Key, V*).

   3.2 *temp = temp ∥ V.*

4. *pseudorandom_bits* = Leftmost (*requested_no_of_bits*) of *temp*.

5. (*Key, V*) = **HMAC_DRBG_Update** (*Null, Key, V*).

6. *reseed_counter = reseed_counter + 1.*

7. **Return** ("Success", *pseudorandom_bits, V, Key, reseed_counter*).

D.3 CTR_DRBG Example Using a Derivation Function

This example of **CTR_DRBG** uses AES-128. The reseed and prediction resistance capabilities are supported, and prediction resistance is obtained during every **Get_entropy_input** call and reseed request. Although the *prediction_resistance_request* parameter in the **Get_entropy_input** and reseed request could be omitted, in this case, they are shown in the pseudocode as a reminder that prediction_resistance will be performed. A block cipher derivation function using AES-128 is used, and a personalization string and additional input are supported. A total of five internal states are available. For this implementation, the functions and algorithms are written as separate routines. **AES_ECB_Encrypt** is the **Block_Encrypt** function (specified in Section 10.4.3) that uses AES-128 in the ECB mode.

The nonce for instantiation (*instantiation_nonce*) consists of a 32-bit incrementing counter. The nonce is initialized when the DRBG is instantiated (e.g., by a call to the clock or by setting it to a fixed value) and is incremented for each instantiation.

The internal state contains the values for *V*, *Key*, *reseed_counter*, and *security_strength*, where *V* and *Key* are bitstrings, and all other values are integers. Since prediction resistance is known to be supported, there is no need for *prediction_resistance_flag* in the internal state.

In accordance with Table 3 in Section 10.2.1, security strengths of 112 and 128 bits may be supported. Using AES-128, the following definitions are applicable for the instantiate, reseed and generate functions:

1. *highest_supported_security_strength* = 128.

2. Output block length (*outlen*) = 128 bits.

3. Key length (*keylen*) = 128 bits.

4. Required minimum entropy for the entropy input during instantiation and reseeding = *security_strength*.

5. Minimum entropy input length (*min _length*) = *security_strength* bits.

6. Maximum entropy input length (*max _length*) = 1000 bits.

7. Maximum personalization string input length (*max_personalization_string_input_length*) = 800 bits.

8. Maximum additional input length (*max_additional_input_length*) = 800 bits.

9. Seed length (*seedlen*) = 256 bits.

10. Maximum number of bits per request (*max_number_of_bits_per_request*) = 4000 bits.

11. Reseed interval (*reseed_interval*) = 100,000 requests.

D.3.1   The CTR_DRBG_Update Function

CTR_DRBG_Update:

    Input: bitstring (*provided_data*, *Key*, *V*).

    Output: bitstring (*Key*, *V*).

    Process:

        1. *temp = Null.*

        2. While (**len** (*temp*) < 256) do

            2.1 $V = (V + 1) \bmod 2^{128}$.

            2.2   *output_block* = **AES_ECB_Encrypt** (*Key*, *V*).

            2.3   *temp = temp* || *ouput_block.*

        3. *temp* = Leftmost 256 bits of *temp.*

        4   *temp = temp* $\oplus$ *provided_data.*

5. *Key* = Leftmost 128 bits of *temp*.

6. *V* = Rightmost 128 bits of *temp*.

7. **Return** (*Key*, *V*).

D.3.2   Instantiation of CTR_DRBG Using a Derivation Function

This implementation will return a text message and an invalid state handle (−1) when an error is encountered. **Block_Cipher_df** is the derivation function in Section 10.4.2, and uses AES-128 in the ECB mode as the **Block_Encrypt** function.

Note that this implementation does not include the *prediction_resistance_flag* in the input parameters, nor save it in the internal state, since prediction resistance is known to be supported.

CTR_DRBG_Instantiate_function:

   Input: integer (*requested_instantiation_security_strength*), bitstring
               personalization_string.

   Output: string *status,* integer *state_handle*.

   Process:

                              Comment: Check the validity of the input
                              parameters.

   1. If (*requested_instantiation_security_strength* > 128) then **Return** ("Invalid *requested_instantiation_security_strength*", −1).

   2. If (**len** (*personalization_string*) > 800), then **Return** ("*Personalization_string* too long", −1).

   3. If (*requested_instantiation_security_strength* ≤ 112), then *security_strength* = 112

      Else *security_strength* = 128.

                              Comment: Get the entropy input.

   4. (*status, entropy_input*) = **Get_entropy_input** (*security_strength, security_strength*, 1000, *prediction_resistance_request*).

   5. If (*status* ≠ "Success"), then **Return** ("Catastrophic failure of the entropy source" || *status*, −1).

                              Comment: Increment the nonce; actual coding
                              must ensure that the nonce wraps when its
                              storage limit is reached, and that the counter
                              pertains to all instantiations, not just this one.

   6. *instantiation_nonce* = *instantiation_nonce* + 1.

                              Comment: Invoke the instantiate algorithm.

7. (*V*, *Key*, *reseed_counter*) = **CTR_DRBG_Instantiate_algorithm** (*entropy_input*, *instantiation_nonce*, *personalization_string*).

> Comment: Find an available internal state and save the initial values.

8. (*status*, *state_handle*) = **Find_state_space** ( ).

9. If (*status* ≠ "Success"), then **Return** ("No available state space:" || *status*, −1).

10. Save the internal state.

> 10.1 *internal_state_* (*state_handle*).V = *V*.
>
> 10.2 *internal_state_* (*state_handle*).Key = *Key*.
>
> 10.3 *internal_state_* (*state_handle*).reseed_counter = *reseed_counter*.
>
> 10.4 *internal_state_* (*state_handle*).security_strength = *security_strength*.

11. **Return** ("Success", *state_handle*).

CTR_DRBG_Instantiate_algorithm:

Input: bitstring (*entropy_input*, *nonce*, *personalization_string*).

Output: bitstring (*V*, *Key*), integer (*reseed_counter*).

Process:

1. *seed_material* = *entropy_input* || *nonce* || *personalization_string*.

2. *seed_material* = **Block_Cipher_df** (*seed_material*, 256).

3. *Key* = 0^{128}.   Comment: 128 bits.

4. *V* = 0^{128}.   Comment: 128 bits.

5. (*Key*, *V*) = **CTR_DRBG_Update** (*seed_material*, *Key*, *V*).

6. *reseed_counter* = 1.

7. **Return** (*V*, *Key*, *reseed_counter*).

D.3.3   Reseeding a CTR_DRBG Instantiation Using a Derivation Function

The implementation is designed to return a text message as the *status* when an error is encountered.

CTR_DRBG_Reseed_function:

Input: integer (*state_handle*), integer *prediction_resistance_request*, bitstring *additional_input*.

Output: string *status*.

Process:

> Comment: Check for the validity of *state_handle*.

1. If ((*state_handle* < 0) or (*state_handle* > 4) or (*internal_state* (*state_handle*) = {*Null, Null*, 0, 0}), then **Return** ("State not available for the indicated *state_handle*").

2. Get the internal state values.

   2.1 *V = internal_state* (*state_handle*).*V*.

   2.2 *Key = internal_state* (*state_handle*).*Key*.

   2.3 *security_strength = internal_state* (*state_handle*).*security_strength*.

3. If (**len** (*additional_input*) > 800), then **Return** ("*additional_input* too long").

4. (*status, entropy_input*) = **Get_entropy_input** (*security_strength, security_strength*, 1000, *prediction_resistance_request*).

6. If (*status* ≠ "Success"), then **Return** ("Catastrophic failure of the entropy source:" || *status*).

                                          Comment: Invoke the reseed algorithm.

7. (*V, Key, reseed_counter*) = **CTR_DRBG_Reseed_algorithm** (*V, Key, reseed_counter, entropy_input, additional_input*).

8. Save the internal state.

   8.1 *internal_state* (*state_handle*). *V = V*.

   8.2 *internal_state* (*state_handle*). *Key = Key*.

   8.3 *internal_state* (*state_handle*). *reseed_counter = reseed_counter*.

   8.4 *internal_state* (*state_handle*). *security_strength = security_strength*.

9. **Return** ("Success").

CTR_DRBG_Reseed_algorithm:

Input: bitstring (*V, Key*), integer (*reseed_counter*), bitstring (*entropy_input, additional_input*).

Output: bitstring (*V, Key*), integer (*reseed_counter*).

Process:

1. *seed_material = entropy_input* || *additional_input*.

2. *seed_material* = **Block_Cipher_df** (*seed_material*, 256).

3. (*Key, V*) = **CTR_DRBG_Update** (*seed_material, Key, V*).

4. *reseed_counter* = 1.

5. **Return** *V, Key, reseed_counter*).

D.3.4   Generating Pseudorandom Bits Using CTR_DRBG

The implementation returns a *Null* string as the pseudorandom bits if an error has been detected.

CTR_DRBG_Generate_function:

> **Input:** integer (*state_handle*, *requested_no_of_bits*, *requested_security_strength*, *prediction_resistance_request*), bitstring *additional_input*.

> **Output:** string *status*, bitstring *pseudorandom_bits*.

> **Process:**

> > Comment: Check the validity of *state_handle*.

> 1. If ((*state_handle* < 0) or (*state_handle* > 4) or (*internal_state* (*state_handle*) = {*Null*, *Null*, 0, 0}), then **Return** ("State not available for the indicated *state_handle*", *Null*).

> 2. Get the internal state.

> > 2.1   *V* = *internal_state* (*state_handle*).*V*.

> > 2.2   *Key* = *internal_state* (*state_handle*).*Key*.

> > 2.3   *security_strength* = *internal_state* (*state_handle*).*security_strength*.

> > 2.4   *reseed_counter* = *internal_state* (*state_handle*).*reseed_counter*.

> > > Comment: Check the rest of the input parameters.

> 3. If (*requested_no_of_bits* > 4000), then **Return** ("Too many bits requested", *Null*).

> 4. If (*requested_security_strength* > *security_strength*), then **Return** ("Invalid *requested_security_strength*", *Null*).

> 5. If (**len** (*additional_input*) > 800), then **Return** ("*additional_input* too long", *Null*).

> 6. *reseed_required_flag* = 0.

> 7. If ((*reseed_required_flag* = 1) OR (*prediction_resistance_flag* = 1)), then

> > 7.1   *status* = **CTR_DRBG_Reseed_function** (*state_handle*, *prediction_resistance_request*, *additional_input*).

> > 7.2   If (*status* ≠ "Success"), then **Return** (*status*, *Null*).

> > 7.3   Get the new working state values; the administrative information was not affected.

> > > 7.3.1   *V* = *internal_state* (*state_handle*).*V*.

> > > 7.3.2   *Key* = *internal_state* (*state_handle*).*Key*.

7.3.3   *reseed_counter = internal_state (state_handle).reseed_counter.*

7.4   *additional_input = Null.*

7.5   *reseed_required_flag* = 0.

> Comment: Generate bits using the generate algorithm.

8. (*status, pseudorandom_bits, V, Key, reseed_counter*) = **CTR_DRBG_Generate_algorithm** (*V, Key, reseed_counter, requested_number_of_bits, additional_input*).

9. If (*status* = "Reseed required"), then

    9.1   *reseed_required_flag* = 1.

    9.2   Go to step 7.

10. Update the internal state.

    10.1 *internal_state (state_handle).V = V.*

    10.2 *internal_state (state_handle).Key = Key.*

    10.3 *internal_state (state_handle).reseed_counter = reseed_counter.*

    10.4 *internal_state (state_handle).security_strength = security_strength.*

11. **Return** ("Success", *pseudorandom_bits*).

CTR_DRBG_Generate_algorithm:

Input: bitstring (*V, Key*), integer (*reseed_counter, requested_number_of_bits*) bitstring *additional_input.*

Output: string *status*, bitstring (*returned_bits, V, Key*), integer *reseed_counter.*

Process:

1. If (*reseed_counter* > 100,000), then **Return** ("Reseed required", *Null, V, Key, reseed_counter*).

2. If (*additional_input* ≠ *Null*), then

    2.1   *additional_input* = **Block_Cipher_df** (*additional_input*, 256).

    2.2   (*Key, V*) = **CTR_DRBG_Update** (*additional_input, Key, V*).

    Else *additional_input* $= 0^{256}$.

3. *temp = Null.*

4. While (**len** (*temp*) < *requested_number_of_bits*) do:

    4.1 $V = (V + 1) \bmod 2^{128}$.

    4.2   *output_block* = **AES_ECB_Encrypt** (*Key, V*).

    4.3   *temp = temp || ouput_block.*

5. *returned_bits* = Leftmost (*requested_number_of_bits*) of *temp*.

6. (*Key*, *V*) = **CTR_DRBG_Update** (*additional_input*, *Key*, *V*)

7. *reseed_counter* = *reseed_counter* + 1.

8. **Return** ("Success", *returned_bits*, *V*, *Key*, *reseed_counter*).

D.4 CTR_DRBG Example Without a Derivation Function

This example of **CTR_DRBG** is the same as the previous example except that a derivation function is not used (i.e., full entropy is always available). As in Appendix D.3, the **CTR_DRBG** uses AES-128. The reseed and prediction resistance capabilities are available. Both a personalization string and additional input are supported. A total of five internal states are available. For this implementation, the functions and algorithms are written as separate routines. **AES_ECB_Encrypt** is the **Block_Encrypt** function (specified in Section 10.4.3) that uses AES-128 in the ECB mode.

The nonce for instantiation (*instantiation_nonce*) consists of a 32-bit incrementing counter that is the leftmost bits of the personalization string (Section 8.6.1 states that when a derivation function is used, the nonce, if used, is contained in the personalization string). The nonce is initialized when the DRBG is instantiated (e.g., by a call to the clock or by setting it to a fixed value) and is incremented for each instantiation.

The internal state contains the values for *V*, *Key*, *reseed_counter*, and *security_strength*, where *V* and *Key* are strings, and all other values are integers.Since prediction resistance is known to be supported, there is no need for *prediction_resistance_flag* in the internal state.

In accordance with Table 3 in Section 10.2.1, security strengths of 112 and 128 bits may be supported. The definitions are the same as those provided in Appendix D.3, except that to be compliant with Table 3, the maximum size of the *personalization_string* is 224 bits in order to accommodate the 32-bits of the *instantiation_nonce* (i.e., **len** (*instantiation_nonce*) + **len** (*personalization_string*) must be ≤ *seedlen*, where *seedlen* = 256 bits). In addition, the maximum size of any *additional_input* is 256 bits (i.e., **len** (*additional_input* ≤ *seedlen*)).

D.4.1   The CTR_DRBG_Update Function

The update function is the same as that provided in Appendix D.3.1.

D.4.2   Instantiation of CTR_DRBG Without a Derivation Function

The instantiate function (**CTR_DRBG_Instantiate_function**) is the same as that provided in Appendix D.3.2, except for the following:

- Step 2 is replaced by:

  If (**len** (*personalization_string*) > 224), then **Return** ("*Personalization_string* too long", −1).

- Step 6 is replaced by :

$instantiation\_nonce = instantiation\_nonce + 1.$

$personalization\_string = instantiation\_nonce \| personalization\_string.$

The instantiate algorithm (**CTR_DRBG_Instantiate_algorithm**) is the same as that provided in Appendix D.3.2, except that steps 1 and 2 are replaced by:

$temp = \textbf{len}\ (personalization\_string).$

If $(temp < 256)$, then $personalization\_string = personalization\_string \| 0^{256\text{-}temp}.$

$seed\_material = entropy\_input \oplus personalization\_string.$

D.4.3   Reseeding a CTR_DRBG Instantiation Without a Derivation Function

The reseed function (**CTR_DRBG_Reseed_function**) is the same as that provided in Appendix D.3.3, except that step 3 is replaced by:

If $(\textbf{len}\ (additional\_input) > 256)$, then **Return** (*"additional_input too long"*).

The reseed algorithm (**CTR_DRBG_Reseed_algorithm**) is the same as that provided in Appendix D.3.3, except that steps 1 and 2 are replaced by:

$temp = \textbf{len}\ (additional\_input).$

If $(temp < 256)$, then $additional\_input = additional\_input \| 0^{256\text{-}temp}.$

$seed\_material = entropy\_input \oplus additional\_input.$

D.4.4   Generating Pseudorandom Bits Using CTR_DRBG

The generate function (**CTR_DRBG_Generate_function**) is the same as that provided in Appendix D.3.4, except that step 5 is replaced by :

If $(\textbf{len}\ (additional\_input) > 256)$, then **Return** (*"additional_input too long"*, *Null*).

The generate algorithm (**CTR_DRBG_Generate_algorithm**) is the same as that provided in Appendix D.3.4, except that step 2.1 is replaced by:

$temp = \textbf{len}\ (additional\_input).$

If $(temp < 256)$, then $additional\_input = additional\_input \| 0^{256\text{-}temp}.$

D.5 Dual_EC_DRBG Example

This example of **Dual_EC_DRBG** allows a consuming application to instantiate using any of the three prime curves. The elliptic curve to be used is selected during instantiation in accordance with the following:

requested_instantiation_security_strength	**Elliptic Curve**
≤ 112	P-256
$113 - 128$	P-256

129 – 192	P-384
193 – 256	P-521

The reseed and prediction resistance capabilities are not supported. Both a *personalization_string* and an *additional_input* are allowed. A total of ten internal states are provided. For this implementation, the algorithms are provided as inline code within the functions.

The nonce for instantiation (*instantiation_nonce*) consists of a random value with *security_strength*/2 bits of entropy; the nonce is obtained by a separate call to the **Get_entropy_input** routine than that used to obtain the entropy input itself. Also, the **Get_entropy_input** function uses only two input parameters, since the first two parameters (the *min_entropy* and the *min_length*) have the same value.

In accordance with Table 4 in Section 10.3.1, security strengths of 112, 128, 192 and 256 bits may be supported. SHA-256 has been selected as the hash function. The following definitions are applicable for the instantiate, reseed and generate functions:

1. *highest_supported_security_strength* = 256.

2. Output block length (*outlen*) = *max_outlen*. See Table 4.

3. Required minimum entropy for the entropy input at instantiation and reseed = *security_strength*.

4. Maximum entropy input length (*max _length*) = 1000 bits.

5. Maximum personalization string length (*max_personalization_string_length*) = 800 bits.

6. Maximum additional input length (*max_additional_input_length*) = 800 bits.

7. Seed length (*seedlen*): = 2 × *security_strength*.

8. Maximum number of bits per request (*max_number_of_bits_per_request*) = 1000 bits.

9. Reseed interval (*reseed_interval*) = 2^{32} blocks.

D.5.1   Instantiation of Dual_EC_DRBG

This implementation will return a text message and an invalid state handle (−1) when an **ERROR** is encountered. **Hash_df** is specified in Section 10.4.1.

Dual_EC_DRBG_Instantiate_function:

> **Input:** integer (*requested_instantiation_security_strength*), bitstring *personalization_string*.

> **Output:** string *status*, integer *state_handle*.

> **Process:**

Comment : Check the validity of the input parameters.

1. If (*requested_instantiation_security_strength* > 256) then **Return** ("Invalid *requested_instantiation_security_strength*", −1).

2. If (**len** (*personalization_string*) > 800), then **Return** ("*personalization_string* too long", −1).

Comment : Select the prime field curve in accordance with the *requested_instantiation_security_strength*.

3. If *requested_instantiation_security_strength* ≤ 112), then

   {*security_strength* = 112; *seedlen* = 224; *outlen* = 240}

   Else if (*requested_instantiation_security_strength* ≤ 128), then

   {*security_strength* = 128; *seedlen* = 256; *outlen* = 240}

   Else if (*requested_instantiation_security_strength* ≤ 192), then

   {*security_strength* = 192; *seedlen* = 384; *outlen* = 368}

   Else {*security_strength* = 256; *seedlen* = 512; *outlen* = 504}.

Comment: Request *entropy_input*.

4. (*status*, *entropy_input*) = **Get_entropy_input** (*security_strength*, 1000).

5. If (*status* ≠ "Success"), then **Return** ("Catastrophic failure of the *entropy_input* source:" ‖ *status*, −1).

6. (*status*, *instantiation_nonce*) = **Get_entropy_input** (*security_strength*/2, 1000).

7. If (*status* ≠ "Success"), then **Return** ("Catastrophic failure of the random nonce source:" ‖ *status*, −1).

Comment: Perform the instantiate algorithm.

8. *seed_material* = *entropy_input* ‖ *instantiation_nonce* ‖ *personalization_string*.

9. *s* = **Hash_df** (*seed_material*, *seedlen*).

10. *reseed_counter* = 0.

11. Using the *security_strength* and the table in Appendix D.5, obtain the domain parameters p, a, b, n, P, and Q from the appropriate elliptic curve.

Comment: Find an unused internal state and save the initial values.

12. (*status*, *state_handle*) = **Find_state_space** ( ).

13. If (*status* ≠ "Success"), then **Return** (*status*, −1).

14. Save the internal state.

12.1   *internal_state (state_handle).s = s.*

14.2   *internal_state (state_handle).seedlen = seedlen.*

14.3   *internal_state (state_handle).p = p.*

14.4   *internal_state (state_handle).a = a.*

14.5   *internal_state (state_handle).b = b.*

14.6   *internal_state (state_handle).n = n.*

14.7   *internal_state (state_handle).P = P.*

14.8   *internal_state (state_handle).Q = Q.*

14.9   *internal_state (state_handle).reseed_counter = reseed_counter.*

14.10 *internal_state (state_handle).security_strength = security_strength.*

15. **Return** ("Success", *state_handle*).

D.5.2 Generating Pseudorandom Bits Using Dual_EC_DRBG

The implemenation returns a *Null* string as the pseudorandom bits if an error is encountered.

Dual_EC_DRBG_Generate_function:

Input: integer (*state_handle, requested_security_strength, requested_no_of_bits*), bitstring *additional_input*.

Output: string *status,* bitstring *pseudorandom_bits*.

Process:

Comment: Check for an invalid *state_handle*.

1. If ((*state_handle* < 0) or (*state_handle* > 9) or (*internal_state (state_handle)* = 0)), then **Return** ("State not available for the *state_handle*", *Null*).

2. Get the appropriate *state* values for the indicated *state_handle*.

2.1   *s = internal_state (state_handle).s.*

2.2   *seedlen = internal_state (state_handle).seedlen.*

2.3   *P = internal_state (state_handle).P.*

2.4   *Q = internal_state (state_handle).Q.*

2.5   *security_strength = internal_state (state_handle).security_strength.*

2.6   *reseed_counter = internal_state (state_handle).reseed_counter.*

Comment: Check the rest of the input parameters.

3. If (*requested_number_of_bits* > 1000), then **Return** ("Too many bits requested", *Null*).

4. If (*requested_security_strength* > *security_strength*), then **Return** ("Invalid requested_strength", *Null*).

5. If (**len** (*additional_input*) > 800), then **Return** ("*additional_input* too long", *Null*).

> Comment: Check whether a reseed is required.

6. If $\left(reseed\_counter + \left\lceil \dfrac{requested\_number\_of\_bits}{outlen} \right\rceil > 2^{32}\right)$, then Return ("DRBG can no longer be used. A new instantiation is required ", *Null*).

> Comment: Execute the generate algorithm.

7. If (*additional_input* = *Null*) then *additional_input* = 0

> Comment: *additional_input* set to *m* zeroes.

Else *additional_input* = **Hash_df** (**pad8** (*additional_input*), *seedlen*).

> Comment: Produce *requested_no_of_bits*, *outlen* bits at a time:

8. *temp* = the *Null* string.

9. *i* = 0.

10. *t* = *s* ⊕ *additional_input*.

11. $s = \varphi(x(t * P))$.

12. $r = \varphi(x(s * Q))$.

13. *temp* = *temp* || (**rightmost** *outlen* bits of *r*).

14. *additional_input* = $0^{seedlen}$.   Comment: *seedlen* zeroes; *additional_input* is added only on the first iteration.

15. *i* = *i* + 1.

16. If (**len** (*temp*) < *requested_no_of_bits*), then go to step 10.

17. *pseudorandom_bits* = **Truncate** (*temp*, *i* × *outlen*, *requested_no_of_bits*).

18. Update the changed value in the *state*.

    18.1   *internal_state.s* = $\varphi(x(s * P))$.

    18.2   *internal_state.reseed_counter* = *reseed_counter*.

19. **Return** ("Success", *pseudorandom_bits*).

Appendix E: (Informative) DRBG Mechanism Selection

Almost no application or system designer starts with the primary purpose of generating good random bits. Instead, the designer typically starts with a goal that he wishes to accomplish, then decides on cryptographic mechanisms, such as digital signatures or block ciphers that can help him achieve that goal. Typically, as the requirements of those cryptographic mechanisms are better understood, he learns that random bits will need to be generated, and that this must be done with great care so that the cryptographic mechanisms will not be weakened. At this point, there are three things that may guide the designer's choice of a DRBG mechanism:

a. He may already have decided to include a set of cryptographic primitives as part of his implementation. By choosing a DRBG mechanism based on one of these primitives, he can minimize the cost of adding that DRBG mechanism. In hardware, this translates to lower gate count, less power consumption, and less hardware that must be protected against probing and power analysis. In software, this translates to fewer lines of code to write, test, and validate.

   For example, a module that generates RSA signatures has an available hash function, so a hash-based DRBG mechanism (e.g., **Hash_DRBG** or **HMAC_DRBG**) is a natural choice.

b. He may already have decided to trust a block cipher, hash function, keyed hash function, etc., to have certain properties. By choosing a DRBG mechanism based on similar properties, he can minimize the number of algorithms he has to trust.

   For example, an AES-based DRBG mechanism (i.e., **CTR_DRBG** using AES) might be a good choice when a module provides encryption with AES. Since the security of the module is dependent on the strength of AES, the module's security is not made dependent on any additional cryptographic primitives or assumptions.

c. Multiple cryptographic primitives may be available within the system or consuming application, but there may be restrictions that need to be addressed (e.g., code size or performance requirements).

   For example, a module with support for both hash functions and block ciphers might use the **CTR_DRBG** if the ability to parallize the generation of random bits is needed.

The DRBG mechanisms specified in this Recommendation have different performance characteristics, implementation issues, and security assumptions.

E.1 Hash_DRBG

Hash_DRBG is based on the use of an **approved** hash function in a counter mode similar to the counter mode specified in NIST [SP 800-38A]. For each generate request, the current value of V (a secret value in the internal state) is used as the starting counter that is iteratively changed to generate each successive *outlen*-bit block of requested output, where *outlen* is the number of bits in the hash function output block. At the end of the generate request, and before the pseudorandom output is returned to the consuming application, the secret value V is updated in order to prevent backtracking.

Performance. The **Generate function** is parallelizable, since it uses the counter mode. Within a generate request, each *outlen*-bit block of output requires one hash function computation and several addition operations; an additional hash function computation is required to provide the backtracking resistance. **Hash_DRBG** produces pseudorandom output bits in about half the time required by **HMAC_DRBG**.

Security. **Hash_DRBG**'s security depends on the underlying hash function's behavior when processing a series of sequential input blocks. If the hash function is replaced by a random oracle, **Hash_DRBG** is secure. It is difficult to relate the properties of the hash function required by **Hash_DRBG** with common properties, such as collision resistance, pre-image resistance, or pseudorandomness. There are known problems with **Hash_DRBG** when the DRBG is instantiated with insufficient entropy for the requested security strength, and then later provided with enough entropy to attain the amount of entropy required for the security strength, via the inclusion of additional input during a generate request. However, these problems do not affect the DRBG's security when **Hash_DRBG** is instantiated with the amount of entropy specified in this Recommendation.

Constraints on Outputs. As shown in Table 2 of Section 10.1, for each hash function, up to 2^{48} generate requests may be made, each of up to 2^{19} bits.

Resources. **Hash_DRBG** requires access to a hash function, and the ability to perform addition with *seedlen*-bit integers. **Hash_DRBG** uses the hash-based derivation function **Hash_df** (specified in Section 10.4.1) during instantiation and reseeding. Any implementation requires the storage space required for the internal state (see Section 10.1.1.1).

Algorithm Choices. The choice of hash functions that may be used by **Hash_DRBG** is discussed in Section 10.1.

E.2 HMAC_DRBG

HMAC_DRBG is built around the use of an **approved** hash function using the HMAC construction. To generate pseudorandom bits from a secret key (*Key*) and a starting value V, the **HMAC_DRBG** computes

$$V - \mathbf{HMAC}\ (Key,\ V).$$

At the end of a generation request, the **HMAC_DRBG** generates a new *Key* and V, each requiring one HMAC computation.

Performance. **HMAC_DRBG** produces pseudorandom outputs considerably more slowly than the underlying hash function processes inputs; for SHA-256, a long generate request produces output bits at about 1/4 of the rate that the hash function can process input bits. Each generate request also involves additional overhead equivalent to processing 2048 extra bits with SHA-256. Note, however, that hash functions are typically quite fast; few if any consuming applications are expected to need output bits faster than **HMAC_DRBG** can provide them.

Security. The security of **HMAC_DRBG** is based on the assumption that an **approved** hash function used in the HMAC construction is a pseudorandom function family. Informally, this means that when an attacker does not know the key used, HMAC outputs look random, even given knowledge and control over the inputs. In general, even relatively weak hash functions seem to be quite strong when used in the HMAC construction. On the other hand, there is not a reduction proof from the hash function's collision resistance properties to the security of the DRBG; the security of **HMAC_DRBG** ultimately relies on the pseudorandomness properties of the underlying hash function. Note that the pseudorandomness of HMAC is a widely used assumption in designs, and the **HMAC_DRBG** requires far less demanding properties of the underlying hash function than **Hash_DRBG**.

Constraints on Outputs. As shown in Table 2 of Section 10.1, for each hash function, up to 2^{48} generate requests may be made, each of up to 2^{19} bits.

Resources. HMAC_DRBG requires access to a dedicated HMAC implementation for optimal performance. However, a general-purpose hash function implementation can always be used to implement HMAC. Any implementation requires the storage space required for the internal state (see Section 10.1.2.1).

Algorithm Choices. The choice of hash functions that may be used by **HMAC_DRBG** is discussed in Section 10.1.

E.3 CTR_DRBG

CTR_DRBG is based on using an **approved** block cipher algorithm in counter mode (see [SP 800-38A]). At the present time, only three-key TDEA and AES are **approved** for use by the Federal government for use in this DRBG mechanism. Pseudorandom outputs are generated by encrypting successive values of a counter; after a generate request, a new key and new starting counter value are generated.

Performance. For large generate requests, **CTR_DRBG** produces outputs at the same speed as the underlying block cipher algorithm encrypts data. Furthermore, **CTR_DRBG** is parallelizeable. At the end of each generate request, work equivalent to two, three or four encryptions is performed, depending on the choice of underlying block cipher algorithm, to generate new keys and counters for the next generate request.

Security. The security of **CTR_DRBG** is directly based on the security of the underlying block cipher algorithm, in the sense that, as long as some limits on the total number of outputs are observed, any attack on **CTR_DRBG** represents an attack on the underlying block cipher algorithm.

Constraints on Outputs. As shown in Table 3 of Section 10.2.1, for each of the three AES key sizes, up to 2^{48} generate requests may be made, each of up to 2^{19} bits, with a negligible chance of any weakness that does not represent a weakness in AES. However, the smaller block size of TDEA imposes more constraints: each generate request is limited to 2^{13} bits, and at most, 2^{32} such requests may be made.

Resources. **CTR_DRBG** may be implemented with or without a derivation function.

When a derivation function is used, **CTR_DRBG** can process the personalization string and any additional input in the same way as any other DRBG mechanism, but at a cost in performance because of the use of the derivation function (as opposed to not using the derivation function; see below). Such an implementation may be seeded by any **approved** source of entropy input that may or may not provide full entropy.

When a derivation function is not used, **CTR_DRBG** is more efficient when the personalization string and any additional input are provided, but is less flexible because the lengths of the personalization string and additional input cannot exceed *seedlen* bits. Such implementations must be seeded by a source of entropy input that provides full entropy (e.g., an **approved** entropy source that has full entropy output or an **approved** NRBG).

CTR_DRBG requires access to a block cipher algorithm, including the ability to change keys, and the storage space required for the internal state (see Section 10.2.1.1).

Algorithm Choices. The choice of block cipher algorithms and key sizes that may be used by **CTR_DRBG** is discussed in Section 10.2.1.

E.4 Dual_EC_DRBG

The **Dual_EC_DRBG** generates pseudorandom outputs by extracting bits from elliptic curve points. The secret, internal state of the DRBG is a value s that is the x-coordinate of a point on an elliptic curve. Outputs are produced by first computing r to be the x-coordinate of the point $s*P$, and then extracting low order bits from the x-coordinate of the elliptic curve point $r*Q$.

Performance. Due to the elliptic curve arithmetic involved in this DRBG mechanism, this algorithm generates pseudorandom bits more slowly than the other DRBG mechanisms in this Recommendation. It should be noted, however, that the design of this algorithm allows for certain performance-enhancing possibilities. First, note that the use of fixed base points allows a substantial increase in the performance of this DRBG mechanism via the use of tables. By storing multiples of the points P and Q, the elliptic curve multiplication can be accomplished via point additions rather than multiplications, a much less expensive operation. In more constrained environments where table storage is not an option, the use of so-called Montgomery Coordinates of the form $(X: Z)$ can be used as a method to increase performance, since the y-coordinates of the computed points are not required. Alternatively, Jacobian or Projective Coordinates of the form (X, Y, Z) can speed up the elliptic curve multiplication operation. These have been shown to be competitive with Montgomery for the NIST-recommended curves in [FIPS 186], and are straightforward to implement.

A given implementation of this DRBG mechanism need not include all three of the **approved** curves in Appendix A. Once the designer decides upon the strength required by a given application, he can then choose to implement the single curve that most appropriately meets this requirement. For a common level of optimization expended, the higher-strength curves will be slower and tend toward less efficient use of output blocks. To mitigate the latter, the designer should be aware that every distinct request for random bits requires the computational expense of at least two elliptic curve point multiplications.

Applications requiring large blocks of random bits (such as IKE or SSL), can thus be implemented most efficiently by first making a single call to the **Dual_EC_DRBG** for all the required bits, and then appropriately partitioning these bits as required by the protocol. For applications that already have hardware or software support for elliptic curve arithmetic, this DRBG mechanism is a natural choice, as it allows the designer to utilize existing capabilities to generate random bits.

Security. The security of **Dual_EC_DRBG** is based on the Elliptic Curve Discrete Logarithm Problem that has no known attacks better than the meet-in-the-middle attacks. For an elliptic curve defined over a field of size 2^m, the work factor of these attacks is approximately $2^{m/2}$, so that solving this problem is computationally infeasible for the curves in this Recommendation. The **Dual_EC_DRBG** is the only DRBG mechanism in this Recommendation whose security is related to a hard problem in number theory.

Constraints on Outputs. For any one of the three elliptic curves listed in Appendix A.1, a particular instance of **Dual_EC_DRBG** may generate at most 2^{32} output blocks before reseeding, where the size of the output blocks is discussed in Section 10.3.1.4. Since the sequence of output blocks is expected to cycle in approximately sqrt(n) bits (where n is the (prime) order of the particular elliptic curve being used), this is quite a conservative reseed interval for any one of the three curves.

Resources. Any source of entropy input may be used with **Dual_EC_DRBG**, provided that it is capable of generating at least *min_entropy* bits of entropy in a string of *max_length* = 2^{13} bits. This DRBG mechanism also requires an appropriate hash function (see Table 4) that is used exclusively for producing an appropriately sized initial state from the entropy input at instantiation or reseeding. An implementation of this DRBG mechanism must also have enough storage for the internal state (see 10.3.1.1). Some optimizations require additional storage for moderate to large tables of pre-computed values.

Algorithm Choices. The choice of appropriate elliptic curves and points used by **Dual_EC_DRBG** is discussed in Appendix A.1.

E.5 Summary for DRBG Selection

Table E-1 provides a summary of the costs and constraints of the DRBG mechanisms in this Recommendation.

Table E-1: DRBG Mechanism Summary

	Dominating Cost/Block	**Constraints (max.)**
Hash_DRBG	2 hash function calls	2^{48} calls of 2^{19} bits
HMAC_DRBG	4 hash function calls	2^{48} calls of 2^{19} bits

CTR_DRBG (TDEA)	1 TDEA encrypt	2^{32} calls of 2^{13} bits
CTR_DRBG (AES)	1 AES encrypt	2^{48} calls of 2^{19} bits
Dual_EC_DRBG	2 EC points	2^{32} blocks

Appendix F : (Informative) Conformance to SP 800-90A Requirements

Conformance to many of the requirements in this Recommendation are the responsibility of entities using, installing or configuring applications or protocols that incorporate the DRBG mechanisms in SP 800-90A, i.e., a given implementation may not have the means of fulfilling these requirements. These requirements include the following:

Section 7.1:

> The entropy input and the seed **shall** be kept secret.
>
> At a minimum, the entropy input **shall** provide the amount of entropy requested by the DRBG mechanism.

Section 7.2:

> The personalization string **shall** be unique for all instantiations of the same DRBG mechanism type (e.g., all instantiations of **HMAC_DRBG**).

Section 8.2:

> A DRBG is instantiated using a seed and may be reseeded; when reseeded, the seed **shall** be different than the seed used for instantiation.

Section 8.3:

> The internal state for an instantiation includes: …One or more values that are derived from the seed and become part of the internal state; these values **shall** remain secret.
>
> The internal state **shall** be protected at least as well as the intended use of the pseudorandom output bits requested by the consuming application.
>
> Each DRBG instantiation **shall** have its own internal state.
>
> The internal state for one DRBG instantiation **shall not** be used as the internal state for a different instantiation.

Section 8.5:

> Within a DRBG mechanism boundary,
>
> 1. The DRBG internal state and the operation of the DRBG mechanism functions **shall** only be affected according to the DRBG mechanism specification.
>
> 2. The DRBG internal state **shall** exist solely within the DRBG mechanism boundary. The internal state **shall not** be accessible by non-DRBG functions or other instantiations of that or other DRBGs.
>
> 3. Information about secret parts of the DRBG internal state and intermediate values in computations involving these secret parts **shall not** affect any information that leaves the DRBG mechanism boundary, except as specified for the DRBG pseudorandom bit outputs.

Other applications may use the same cryptographic primitive, but the DRBG's internal state and the DRBG mechanism functions **shall not** be affected by these other applications.

The boundary around the entire DRBG mechanism **shall** include the aggregation of sub-boundaries providing the DRBG mechanism functionality.

When DRBG mechanism functions are distributed, a secure channel **shall** be used to protect the confidentiality and integrity of the internal state or parts of the internal state that are transferred between the distributed DRBG mechanism sub-boundaries.

The security provided by the secure channel **shall** be consistent with the security required by the consuming application.

Section 8.6.3:

The entropy input **shall** have entropy that is equal to or greater than the security strength of the instantiation.

Section 8.6.5:

The source of the entropy input (SEI) **shall** be either:

1. An **approved** entropy source,

2. An **approved** NRBG (note that an NRBG includes an entropy source), or

3. An **approved** DRBG, thus forming a chain of at least two DRBGs; the initial DRBG in the chain **shall** be seeded by an **approved** NRBG or an **approved** entropy source.

A DRBG instantiation may seed or reseed another DRBG instantiation, but **shall not** reseed itself.

Section 8.6.6:

The entropy input and the resulting seed **shall** be handled in a manner that is consistent with the security required for the data protected by the consuming application. For example, if the DRBG is used to generate keys, then the entropy inputs and seeds used to generate the keys **shall** (at a minimum) be protected as well as the keys.

Section 8.6.7:

The nonce **shall** be either:

a. A value with at least (1/2 *security_strength*) bits of entropy,

b. A value that is expected to repeat no more often than a (1/2 *security_strength*)-bit random string would be expected to repeat.

Section 8.6.9:

The seed that is used to initialize one instantiation of a DRBG **shall not** be intentionally used to reseed the same instantiation or used as the seed for another DRBG instantiation.

In addition, a DRBG **shall not** reseed itself.

Section 8.6.10:

The seed used by a DRBG and the entropy input used to create that seed **shall not** intentionally be used for other purposes (e.g., domain parameter or prime number generation).

Section 10.2.1:

The use of the derivation function is optional if either an **approved** RBG or an entropy source provide full entropy output when entropy input is requested by the DRBG mechanism. Otherwise, the derivation functon **shall** be used.

Section 11:

Therefore, entropy input used for testing **shall not** knowingly be used for normal operational use.

Appendix G : (Informative) References

[FIPS 140]   Federal Information Processing Standard 140-2, *Security Requirements for Cryptographic Modules*, May 25, 2001.

[FIPS 180]   Federal Information Processing Standard 180-4, *Secure Hash Standard (SHS)*, February 2011 Draft

[FIPS 186]   Federal Information Processing Standard 186-3, *Digital Signature Standard (DSS)*, Draft March 2006.

[FIPS 197]   Federal Information Processing Standard 197, *Advanced Encryption Standard (AES)*, November 2001.

[FIPS 198]   Federal Information Processing Standard 198-1, *Keyed-Hash Message Authentication Code (HMAC)*, July 2008.

[SP 800-38A]   National Institute of Standards and Technology Special Publication (SP) 800-38A, *Recommendation for Block Cipher Modes of Operation - Methods and Techniques*, December 2001.

[SP 800-57]   NIST Special Publication (SP) 800-57, Part 1, *Recommendation for Key Management*: General, [August 2005].

[SP 800-67]   NIST Special Publication (SP) 800-67, *Recommendation for the Triple Data Encryption Algorithm (TDEA) Block Cipher*, May 2004.

[X9.62]   American National Standard (ANS) X9.62-2005, *Public Key Cryptography for the Financial Services Industry - The Elliptic Curve Digital Signature Algorithm (ECDSA)*.

[X9.63]   American National Standard (ANS) X9.63-2001, *Public Key Cryptography for the Financial Services Industry - Key Agreement and Key Transport Using Elliptic Key Cryptography*.

[Gurel]   Gurel, Nicholas, "Extracting Bits from Coordinates of a Point of an Elliptic Curve", Cryptology Eprint Archive: 2005/324.

[Shparlinski]   E.E. Mahassni and I. Shparlinski, On the Uniformity of Distribution of Congruential Generators over Elliptic Curves, preprint November 2000. http://citeseer.ist.psu.edu/mahassni00uniformity.html .

Appendix H : (Informative) Revisions

This original version of this Recommendation was completed in June, 2006. In March 2007, the following changes were made (note that the changes are indicated in italics):

1. Section 8.3, item 1.a originally stated the following:

    "One or more values that are derived from the seed and become part of the internal state; these values must usually remain secret"

    The item now reads:

    "One or more values that are derived from the seed and become part of the internal state; these values *should* remain secret".

2. In Section 8.4, the third sentence originally stated:

    "Any security strength may be requested, but the DRBG will only be instantiated to one of the four security strengths above, depending on the DRBG implementation."

    The sentence now reads:

    "Any security strength may be requested *(up to a maximum of 256 bits)*, but the DRBG will only be instantiated to one of the four security strengths above, depending on the DRBG implementation."

3. In Section 8.7.1, the list of examples of information that could appear in a personalization string included privater keys, PINs and passwords. These items were removed from the list, and seedfiles were added.

4. In Section 10.3.1.4, a step was inserted that will provide backtracking resistance (step 14 of the pseudocode). The same change was made to the example in Appendix D.5.3 (step 19.1). In addition, the two occurrences of *block_counter* (in input 1 and processing step 1) were corrected to be *reseed_counter*.

This Recommendation was developed in concert with American National Standard (ANS) X9.82, a multi-part standard on random number generation. Many of the DRBGs in this Recommendation and the requirements for using and validating them are also provided in ANS X9.82, Part 3. Other parts of that Standard discuss entropy sources and RBG construction. During the development of the latter two documents, the need for additional requirements and capabilities for DRBGs were identified. As a result, the following changes were made to this Recommendation in August 2008 :

1. Definitions have been added in Section 4 for the following: **approved** entropy source, DRBG mechanism, fresh entropy, ideal random bitstring, ideal random sequence and secure channel. The following definitions have been modified: backtracking resistance, deterministic random bit generator (DRBG), entropy, entropy input, entropy source, full entropy, min-entropy, prediction resistance, reseed, security strength, seed period and source of entropy input.

2. In Section 6, a link was provided to examples for the DRBGs specified in this Recommendation.

3. In Section 7.2, paragraph 3. 2nd sentence: The "**should**" has been changed to "**shall**", so that the sentence now reads:

    The personalization string *shall* be unique for all instantiations of the same DRBG mechanism type (e.g., **HMAC_DRBG**).

4. In Section 8.2, paragraph 2, additional text was added to the the first sentence, which now reads:

    A DRBG is instantiated using a seed and may be reseeded; *when reseeded, the seed **shall** be different than the seed used for instantiation.*

5. In Section 8.5, Figure 4 has been updated, and the last paragraph has been revised to discuss the use of a secure channel.

6. In Sections 8.6.5 and 8.6.9, statements were inserted that prohibit a DRBG instantiation from reeeding itself.

7. References to "entropy input" have been removed from Section 8.6.9.

8. Section 8.8: An example was added to further clarify the meaning of prediction resistance.

9. In Section 9, a *prediction_resistance_request* parameter has been added to the **Get_entropy_input** call, along with a description of its purpose to the text underneath the call.

10. In Section 9, a footnote was inserted to explain why a *prediction_resistance_requst* parameter may be useful in the **Get_entropy_input** call.

11. In Section 9.1, the following changes were made:

    * The following sentence has been added to the description of the *prediction_resistance_flag*:

      In addition, step 6 can be modified to not perform a check for the *prediction_resistance_flag* when the flag is not used in an implementation ; in this case, the **Get_entropy_input** call need not include the *prediction_resistance_request* parameter.

    * The following requirement has been added to the **Required information not provided by the consuming application during instantiation.**

      This input **shall not** be provided by the consuming application as an input parameter during the instantiate request.

    * A *prediction_resistance_request* parameter has been added to the **Get_entropy_input** call of step 6 of the **Instantiate Process**.

- Step 5 was originally intended for implementations of the **Dual_EC_DRBG** to select an appropriate curve. This function is now performed by the **Dual_EC_DRBG**'s Instantiate_algorithm. Changes were made to provide the security strength to the Instantiate_algorithm. The Instantiate_algortihm for each DRBG was changed to allow the input of the security strength.

12. In Section 9.2, the following changes have been made:

    - A *prediction_resistance_request* parameter has been added to the **Reseed_function** call.

    - A description of the parameter has been added below the function call.

    - A step was inserted that checked a request for prediction resistance (via the *prediction_resistance_request* parameter) against the state of the *prediction_resistance_flag* that may have been set during instantiation.

    - A *prediction_resistance_request* parameter has been added to the **Get_entropy_input** call of (newly numbered) step 4 of the **Reseed Process**.

    - In the description of the *entropy_input* parameter, a restriction was added that the *entropy_input* is not to be provided by the instantiation being reseeded. by the DRBG instantiation being reseeded.

    - A footnote was inserted to explain why the prediction_resistance_request parameter might be useful.

13. In Section 9.3.1, the following changes were made:

    - Text has been added to item to refer to the **Reseed_function**.

    - A *prediction_resistance_request* parameter has been added to the **Get_entropy_input** call of step 7.1 of the **Generate Process**.

    - A substep was inserted in step 9 of the **Generate Process** to check the *prediction_resistance request* against the state of the *prediction_resistance_flag*.

14. In Section 9.3.2, step e, a phrase addressing the presence of the *prediction_resistance_request* indicator was inserted.

15. In Sections 10.1 and 10.3.1, the new hash functions approved in FIPS 180-4 have been added.

16. In Sections 10.1.2 (**HMAC_DRBG**) and 10.2.1 (**CTR_DRBG**), the update functions have been renamed to reflect the DRBG with which they are associated (i.e., renamed ro **HMAC_DRBG_Update** and **CTR_DRBG_Update**).

17. In Section 10.1.2.1, the last paragraph has been revised to indicate that only the Key is considered to be a critical value.

18. In Sections 10.1.2.3, 10.2.1.3.1, 10.2.1.3.2 and 10.3.1.2, the description of the *personalization_string* has been revised to indicate that the length the *personalization_string* may be zero.

19. In Section 10.2.1.5, the following statement has been added to the first paragraph:

    If the derivation function is not used, then the maximum allowed length of *additional_input = seedlen*.

20. In Section 10.3.1.2, the specification was changed to select an elliptic curve and return the parameters of that curve to the **Instantiate_function** that called the routine.

21. In the first paragraph of Appendix A.1, a statement has been added that if alternative points are desired, they **shall** be generated as specified in Appendix A.2.

22. The original Appendices C and D on entropy sources and RBG constructions, respectively, have been removed and the topics will be discussed in SP 800-90B and C

23. In Appendix C.2 (originally Appendix E.2), a paragraph has been inserted after the table of E values that discusses the analysis associated with the table values.

24. The additional uses of the *prediction_resistance_request* parameter (as specified in Section 9) have been added to the following appendices:

    - D.1.1, step 4;
    - D.1.2, Input and step 4;
    - D.1.3, step 7.1;
    - D.3.2, step 4;
    - D.3.3, Input and step 4; and
    - D.3.4, step 7.1.

25. The name of the update call has been changed in the following appendices:

    - D.2.1, step 4;
    - D.2.2, step 5;
    - D.3.1, title; and
    - D.4.1, title.

26. In Appendix D.3 (originally Appendix F.3), the first paragraph, which discusses the example, has been modified to discuss the *prediction_resistance_request* parameter in the **Get_entropy_input** call.

27. In Appendix D.5 (originally Appendix F.5), the description of the example in paragraph 2 has been changed so that the example does not include prediction resistance, and the definition for the *reseed_interval* has been removed from the

list. The **Dual_EC_Instantiate_function** has been modified to reflect the changes made to the **Instantiate_function** and **Instantiate_algorithm** (see the last bullet of modification 8 above). In addition, the pseudocode for the **Reseed_function** has been removed, and steps in F.5.1 and F.5.2 that dealt with reseeding have been removed.